Christian Oberdanner

ROS and Antioxidant Systems in Apoptosis

.

Christian Oberdanner

ROS and Antioxidant Systems in Apoptosis

Oxidant Balance in Cell Death and Cancer
Therapy

VDM Verlag Dr. Müller

Imprint

Bibliographic information by the German National Library: The German National Library lists this publication at the German National Bibliography; detailed bibliographic information is available on the Internet at http://dnb.d-nb.de.

Cover image: www.purestockx.com

Publisher:
VDM Verlag Dr. Müller Aktiengesellschaft & Co. KG , Dudweiler Landstr. 125 a, 66123 Saarbrücken, Germany,
Phone +49 681 9100-698, Fax +49 681 9100-988,
Email: info@vdm-verlag.de

Zugl.: Salzburg, Naturwissenschaftliche Fakultät der Universität Salzburg, Diss., 2007

Produced in USA and UK by:
Lightning Source Inc., La Vergne, Tennessee, USA
Lightning Source UK Ltd., Milton Keynes, UK
BookSurge LLC, 5341 Dorchester Road, Suite 16, North Charleston, SC 29418, USA

ISBN: 978-3-8364-8211-0

Contents

I. Zusammenfassung

Reaktive Sauerstoffspezies (ROS) spielen entscheidende Rollen in diversen biologischen Prozessen. Durch ihre hohe Reaktivität sind ROS hauptsächlich für ihre schädigende Wirkung auf zelluläre Bestandteile wie Lipide, Proteine und DNA bekannt und werden daher mit einer Vielzahl von Krankheiten, insbesondere mit Krebs, assoziiert.

In kontrollierten Konzentrationen sind ROS aber essentiell für lebensnotwendige Prozesse wie den programmierten Zelltod, die Apoptose. In diesem Zusammenhang bewirken ROS mutmaßlich das Öffnen mitochondrialer Poren, und markieren somit den Beginn der mitochondrialen bzw. stressvermittelten Apoptose. Diese Apoptose-induzierende Wirkung von ROS wird in kontrollierter Weise bei verschiedenen Formen der Krebstherapie genutzt, beispielsweise bei der sogenannten Photodynamischen Therapie (PDT), die auf einer auf ein lokales Tumorareal beschränkten Überproduktion von ROS basiert.

Antioxidantien sind die zellulären Gegenspieler reaktiver Sauerstoffradikale. Sie agieren auf unterschiedlichste Weise, um die zytotoxische Wirkung von ROS zu neutralisieren. In einem gesunden Organismus halten intrazelluläre antioxidative Systeme die Konzentration an ROS auf einem Level, der wichtige ROS-mediierte Prozesse, aber keine bzw. nur vernachlässigbare Schäden zuläßt. Eine Verlagerung dieses oxidativen Gleichgewichts auf die eine oder die andere Seite kann schwerwiegende Folgen haben.

Auch in ROS-basierenden Krebstherapien scheint das durch Antioxidantien vermittelte oxidative Gleichgewicht überaus wichtig zu sein. So vermag eine künstlich hervorgerufene verminderte antioxidative Kapazität eine erhöhte zytotoxische und somit therapiefördernde Wirkung hervorzurufen, wogegen die Zugabe von Antioxidantien die Apoptoseinduktion in Tumorzellen und damit das zytotoxische Potential der Therapie verringern kann.

In der vorliegenden Dissertation wurde in drei unterschiedlichen Einzelstudien der Einfluss von veränderter intrazellulärer Antioxidatien-Kapazität auf die Apoptose-induzierende Wirkung von ROS-basierenden Krebstherapien am Modell humaner Plattenepithelkarzinomzellen (A431) untersucht.

In den ersten beiden Studien wurden die Auswirkungen der Modulierung des intrazellulären Glutathion (GSH)-Spiegels auf PDT-induzierte ROS-Produktion und Apoptose untersucht. Aus den Ergebnissen geht hervor, dass durch die Senkung der intrazellulären GSH-Konzentration eine markante Erhöhung der zellulären Sensitivität gegenüber PDT erzielt wird, was auf die Verminderung der essentiellen Antioxidantien-Funktion von GSH zurückzuführen ist.

In der dritten Studie wurde das Apoptose-induzierende Potential des Staurosporin-Analogs Ro-31-8220, das unter anderem bei Chemotherapien eingesetzt wird, untersucht. Die gewonnenen Ergebnisse zeigen, dass Ro-31-8220-vermittelte Apoptose in ursächlichem Zusammenhang mit überhöhter interazellulärer ROS-Produktion steht und überdies durch die Zugabe von Antioxidantien nahezu vollständig verhindert werden kann.

Die erzielten Ergebnisse sind von Relevanz in der klinischen Anwendung, insbesondere bei PDT und bei Staurosporin-gestützten Therapien und tragen darüber hinaus zum allgemeinen Verständnis von ROS/Antioxidant-Interaktionen bei.

I. Abstract

Reactive oxygen species (ROS) are involved in various biological processes. Due to their high reactivity ROS may easily interact with cellular components like lipids, proteins and, most importantly, DNA. Increased intracellular levels of ROS may result in tissue damage and are therefore associated with various diseases, especially with cancer.

However, apart from their potentially harmful functions, ROS have been identified as important mediators of essential cellular processes like programmed cell death. In this context, ROS have been described to induce the opening of mitochondrial permeability transition pores, thus initiating the stress-induced apoptosis pathway. The apoptosis-inducing capacity of ROS is the basis for several different anti-cancer therapies such as photodynamic therapy (PDT) which is based on a local, tumor area-restricted overproduction of ROS.

Antioxidants are the cellular antagonists of ROS and act in many different ways in order to neutralize the cytotoxic impact of oxyradicals. In healthy individuals, antioxidant systems serve to maintain intracellular ROS levels below a certain threshold, permitting the functionality of essential ROS-mediated signaling processes but preventing ROS overproduction and potential tissue damage. Even small shifts of the ROS-antioxidant balance may entail serious biological consequences.

In the context of ROS-based anti-cancer applications, the action of antioxidants may critically influence the outcome of the therapy. Decreased antioxidant capacity that is induced purposefully as a therapeutic strategy favors apoptosis induction and may therefore result in an improved anti-cancer action, whereas the external addition of antioxidants may prevent apoptosis induction and thus reduce the cytotoxic potential of the therapy.

The present doctoral study is composed of three different studies investigating the influence of altered intracellular antioxidant capacity on the apoptosis-inducing potential of several ROS-based forms of tumor therapy, using human epidermoid carcinoma cells (A431) as a model system.

In the first two studies, two different experimental setups were utilized to analyze the impact of the modulation of the intracellular glutathione (GSH) level on PDT-induced ROS production and apoptosis. The results indicate that a decrease of the intracellular GSH level leads to a markedly heightened sensitivity to PDT which can be attributed to the decreased function of the essential antioxidant GSH.

The third study investigated the apoptosis-inducing potential of the staurosporine (STS) analog Ro-31-8220, which is used in various chemotherapies. The obtained results show that Ro-31-8220-mediated apoptosis is causally linked to intracellular ROS overproduction, and that apoptosis induction can be almost completely inhibited by addition of several antioxidants.

Taken together, these observations are of high relevance for various clinical applications, especially for PDT- and staurosporine-mediated therapies, and contribute to the general understanding of ROS/antioxidant interrelations.

II. Introduction

The introduction of this doctoral thesis is divided into two parts.

The first section is on reactive oxygen species (ROS) and antioxidant defense in biological systems.

The second part reviews the general mechanisms of apoptosis and specifically addresses the roles of ROS and antioxidant strategies in the induction of apoptosis.

In addition, the comprehensive review: *"Apoptosis Following Photodynamic Tumor Therapy: Induction, Mechanisms and Detection"* (see appendix), which was published as a central paper during the course of my dissertation provides deeper insight into ROS-mediated apoptosis regulation in the context of Photodynamic Therapy (PDT).

1. Reactive oxygen species and antioxidant defense

In anorganic chemistry, free radicals are defined as atoms or molecules with one or more unpaired electrons and thus possess a high potential to donate electrons to an acceptor molecule. Due to this fact radicals are highly reactive and take part in numerous chemical reactions. They play major roles in combustion chemistry, atmospheric chemistry, plasma chemistry, polymerization chemistry and, above all, in biochemistry.

Free radicals are involved in a number of biological processes some of which are necessary for life, such as the intracellular killing of bacteria by macrophages or certain cell signaling processes. Excess radical concentration may be harmful to the cell because they can easily destroy different cellular components, which may result in the development of serious diseases.

In biological systems the most prominent free radicals are reactive oxygen species (ROS), which are derived from molecular oxygen (O_2). O_2 plays a dual role in an aerobic cell. On the one hand, O_2 is required for cellular respiration as the terminal electron acceptor in the electron transport chain inside mitochondria; on the other hand, O_2 initiates chemical processes which can destroy biological tissue (oxidative stress). A statement made by Skulachev describes this two edged situation most precisely [1].

"Living with the risk of oxidative stress is a price that aerobic organism must pay for more efficient bioenergetics".

1.1. Chemistry of ROS

The term "reactive oxygen species" covers all derivatives of molecular oxygen, those which are themselves highly reactive (like $^{\bullet}OH$) and those which can be easily converted to such radicals (like H_2O_2). Table 1.1. shows the major types of ROS in biological systems, distinguishing between radical and non-radical species.

radical ROS	chemical formula
superoxide radical	$^{\bullet}O_2^-$
hydroxyl radical	$^{\bullet}OH$
peroxide radical	$^{\bullet}OOH$
alkoxyl radical	$RO^{\bullet}$
peroxyl radical	$ROO^{\bullet}$
non-radical ROS	
singlet oxygen	$^1O_2$
hydrogen peroxide	H_2O_2
hydroperoxide	$ROOH$
ozone	O_3

Table 1.1. Major types of ROS in biological systems.

Singlet Oxygen ($^1O_2$)

$^1O_2$ is generated by a chemo-physical activation of oxygen. Due to the fact that the ground state of molecular oxygen has triplet multiplicity ($^3O_2$), $^3O_2$ is not able to react with any organic compound at room temperature, but it is very likely that energy (E) is transferred to it [2, 3]. By the uptake of energy, $^3O_2$ is brought to its activated state, a singlet configuration, termed $^1O_2$.

$$^3O_2 \xrightarrow{\ +E\ } {}^1O_2$$

Reaction 1.1. $^3O_2$ can be easily activated to $^1O_2$ by a simple energy transfer.

$^1O_2$ is very reactive, but due to its very short lifetime (3 µs in water) its major function is to produce other ROS, like $^{\bullet}O_2^-$, rather than to directly interact with tissue [2].

$$^3O_2 \xrightarrow{\quad E \quad} {}^1O_2 \xrightarrow{\quad +e^- \quad} {}^\bullet O_2{}^-$$

Reaction 1.2. $^1O_2$ is converted to $^\bullet O_2^-$ by uptake of a free electron.

Superoxide ($^\bullet O_2^-$)

$^\bullet O_2^-$ is probably the most important ROS in aerobic organisms as it is the one most efficiently produced inside mitochondria. Due to the moderate redox potential of $^\bullet O_2^-/O_2$ (-0.16 V) a one-electron reduction of oxygen is thermodynamically favorable [4].

$$O_2 \xrightarrow{\quad +e^- \quad} {}^\bullet O_2{}^-$$

Reaction 1.3. O_2 is converted to $^\bullet O_2^-$ by a one-electron reduction.

In mitochondria semiquinones ($^\bullet Q^-$) generated in the course of electron transport reactions in the respiratory chain donate electrons to oxygen and provide constant sources of $^\bullet O_2^-$ (see chapter 1.2.)

$$^\bullet Q^- + O_2 \longrightarrow Q + {}^\bullet O_2{}^-$$

Reaction 1.4. Semiquinone donates an electron to O_2 producing $^\bullet O_2^-$ and ubiquinone (Q).

$^\bullet O_2^-$ produced inside eukaryotic cells is usually dismutated very quickly to H_2O_2 by the enzyme superoxide dismutase (SOD) (reaction 1.5.). H_2O_2 can further react to the highly reactive $^\bullet OH$ radical (reactions 1.6. and 1.7.). Therefore, $^\bullet O_2^-$ itself seems not to be the species which produces the most direct and harmful impact on the cell [3].

Hydrogen peroxide (H_2O_2)

H_2O_2 is relatively unreactive since it does not belong to the radical forms of ROS [3]. As mentioned above, most of the H_2O_2 molecules inside the cell are generated by the dismutative action SOD on $^\bullet O_2^-$.

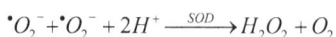

$$^\bullet O_2^- + {}^\bullet O_2^- + 2H^+ \xrightarrow{\quad SOD \quad} H_2O_2 + O_2$$

Reaction 1.5. Dismutation of $^\bullet O_2^-$ to H_2O_2 by cellular SOD.

In contrast to $^\bullet O_2^-$, H_2O_2 is able to penetrate through membranes and may therefore escape from the cell to be diluted in the extracellular space [3, 4]. Moreover, there are mitochondrial and cytosolic antioxidants which can decompose H_2O_2 (see chapter 1.4.), but if these mechanism do not succeed in decomposing all present H_2O_2, it can easily react to form dangerous $^\bullet OH$ radicals.

Hydroxyl Radical ($^\bullet OH$)

$^\bullet OH$ radicals are the most harmful ROS for the eukaryotic cell because of their high redox potential for $^\bullet OH + H^+/H_2O$ (+1.35V), meaning that $^\bullet OH$ can easily oxidize practically everything in the cell [5]. $^\bullet OH$ radicals are mainly produced by two reactions, the Fenton reaction and the Haber-Weiss reaction [3, 6, 7].

In the Fenton reaction (named after H.J.H. Fenton) ferrous iron(II) is oxidized by H_2O_2 to ferrous iron(III), a hydroxyl radical and a hydroxyl anion. Iron(III) is reduced back to iron(II), a peroxide radical and a proton by another hydrogen peroxide molecule (disproportionation).

I) $Fe^{2+} + H_2O_2 \longrightarrow Fe^{3+} + {}^\bullet OH + OH^-$

II) $Fe^{3+} + H_2O_2 \longrightarrow Fe^{2+} + {}^\bullet OOH + H^+$

Reaction 1.6. The Fenton reaction produces dangerous $^\bullet OH$ and $^\bullet OOH$ radicals.

The Haber-Weiss reaction (named after F. Haber and J. Weiss) is based on the findings of H.J.H. Fenton [6, 7]. The Fenton reaction (I) initiates a kind of chain reaction - reactions II and III - while the chain termination is caused by reaction IV.

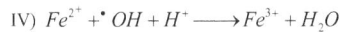

I) $Fe^{2+} + H_2O_2 \longrightarrow Fe^{3+} + {}^\bullet OH + OH^-$ II) $H_2O_2 + {}^\bullet OH \longrightarrow H_2O + {}^\bullet O_2^- + H^+$

III) $^\bullet O_2^- + H_2O_2 + H^+ \longrightarrow O_2 + {}^\bullet OH + H_2O$ IV) $Fe^{2+} + {}^\bullet OH + H^+ \longrightarrow Fe^{3+} + H_2O$

Reaction 1.7. The Haber-Weiss reaction.

The above mentioned ROS are not the only ones inside the eukaryotic cell. There are many, so called *secondary* ROS, which are generated when one of the discussed ROS reacts with cell components in their close spatial proximity, e.g. $ROO^\bullet$, $RO^\bullet$ or $ROOH$. These "organic" ROS have been shown to play major roles in various cellular processes, like in cell signaling [3, 8].

1.2. The generation of ROS in eukaryotic cells

In principle, there are exogenous and endogenous sources for the generation of ROS inside a cell (figure 1.1.). Among harmful exogenous sources are cigarette smoke, ultraviolet and radioactive radiation, certain pesticides or chemicals like asbestos.

Endogenous ROS can be produced in several ways. For example, the endoplasmatic reticulum (ER) contains an enzyme called cytochrome P450. It is known to detoxify several cytotoxic substances like xenobiotics, retinoic acid or eicosanoids, thereby generating high amounts of ROS [3].

Specific cells of the immune system like macrophages produce ROS to act against harmful intruders by using the enzyme NADPH oxidase to generate $^\bullet O_2^-$ [3].

In addition, there are several other enzymes reducing O_2 to $^\bullet O_2^-$ (e.g. xanthine oxidase), but the main origin of ROS inside the eukaryotic cell are mitochondria.

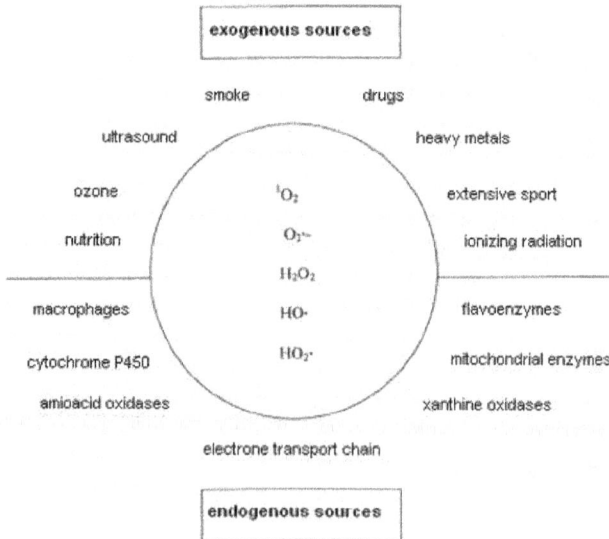

Figure 1.1. Exogenous and endogenous sources for ROS.

- 13 -

The respiration-coupled energy conservation in form of ATP is usually the most important mitochondrial function, providing 90-95% of the total amount of ATP, the rest being synthesized by glycolytic phosphorylation. The adult human forms and decomposes the incredible number of about 40 kg ATP per day.

The mitochondrial enzyme H^+-ATP-synthase catalyzes the respiratory phosphorylation by using the electrochemical H^+ potential difference produced by the respiratory chain (mitochondrial transmembrane potential, $\Delta\psi$) to generate ATP. Different enzyme-complexes in the respiratory chain catalyze the electron transfer from NADH to O_2 to form H_2O (four electron reduction of O_2), providing energy to transfer H^+ from the matrix to the intermembrane space. Due to the fact that H^+ can not diffuse through the membrane, a gradient is formed. The only possibility for H^+ to return into the matrix is via the F_0 part of the ATP-synthase complex, a process which powers the phosphorylation of ADP to ATP (figure 1.2).

This electron transport via the mitochondrial respiratory chain is the place where most intracellular ROS are produced. The respiratory chain production of $^\bullet O_2^-$ is estimated to divert electrons at about 1% of the total rate of electron transport from NADH to oxygen [3, 4].

The respiratory chain is made up of four electron-transporting complexes (C-I to C-IV) and the above mentioned ATP-synthase complex, sometimes referred to as complex V (see figure 1.2.). Two of these complexes are thought to be the major ROS production sites: complex I, the NADH-ubiquinone reductase, and complex III, the ubiquinone cytochrome c reductase.

However, at least nine mammalian mitochondrial enzymes and enzymatic systems could be reported to produce a significant amount of ROS. These nine were referred to as "Nazgul" in a very interesting review by Andreyev and co-workers [9], in allusion to the famous book "The Lord of the Rings" by J.R.R Tolkien [10].

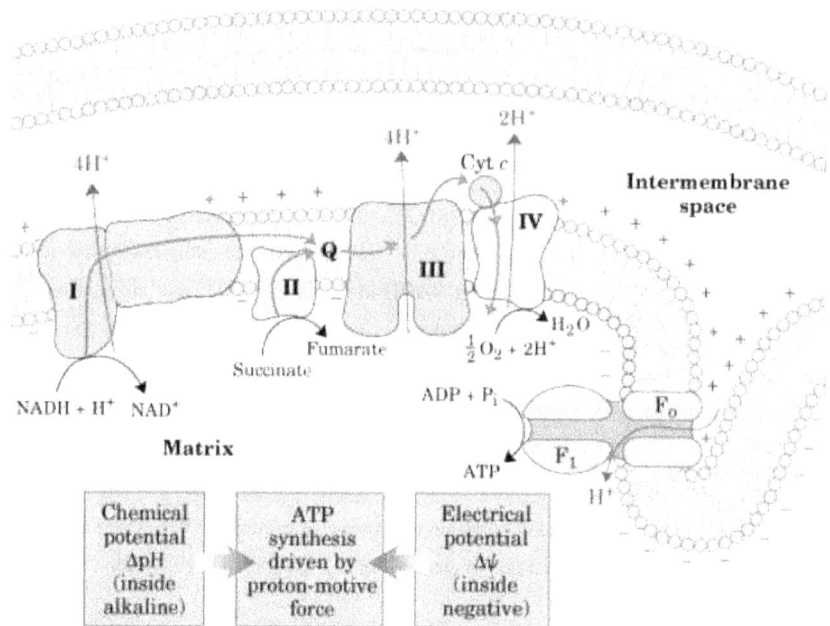

Figure 1.2. The mitochondrial respiratory chain.

C-I, NADH ubiquinone reductase; NADH+H$^+$ reduces FMN$^+$ to FMNH$_2$ which in turn transfers electrons to iron-sulfur clusters and further on to the oxidized form of coenzyme Q (ubiquinone, CoQ or Q), forming reduced coenzyme Q (ubiquinol, CoQH$_2$ or QH$_2$). Four H$^+$ are transferred from the matrix to the intermembrane space.

C-II, succinate ubiquinone reductase; is a part of the citric acid cycle and serves to additionally reduce CoQ molecules but without any proton pump functionality. The electrons derived from the oxidation of succinate to fumarate are transferred via FADH$_2$, iron-sulfur cluster and cytochrome b to CoQ, producing CoQH$_2$.

C-III, ubiquinone cytochrome c reductase; removes in a stepwise fashion two electrons from one CoQH$_2$ molecule and transfers them to two molecules of cytochrome c, again pumping four H$^+$ from the matrix to the intermembrane space.

C-IV, (cytochrome c oxidase); removes four electrons from four molecules cytochrome c and transfers them to O$_2$ in a stepwise manner, producing two molecules H$_2$O. At the same time four H$^+$ are moved across the membrane, producing a proton gradient.

(adapted and from Edward Marcotte, Department of Chemistry & Biochemistry, University of Texas at Austin; http://courses.cm.utexas.edu)

Nine mitochondrial ROS production sites

1. Cytochrome b5 reductase (Cyt.b5 reductase)

Cyt.b5 reductase is an enzyme located in the outer mitochondrial membrane. It oxidizes cytoplasmatic NADH and reduces cytochrome b5. According to Whatley et. al. $^\bullet O_2^-$ is the ROS mainly produced by Cyt.b5 reductase [11].

2. Monoamine oxidase (MAO)

MAO is an enzyme located in the outer mitochondrial membrane which catalyzes the oxidation of biogenic amines accompanied by the release of H_2O_2. MAO is a very efficient producer of H_2O_2 inside mitochondria, maybe the major source for H_2O_2 in tissues with ischemia, aging and during oxidation of exogenous amines [12-14]. MAO is thought to be responsible for the damage of mitochondria in Parkinson´s disease [15].

3. Dihydroorotate dehydrogenase (DHOH)

DHOH is located at the outer surface of the inner mitochondrial membrane and catalyzes the conversion of dihydroorotate to orotate, a reaction performed during the synthesis of pyrimidine nucleotides. Reduced DHOH is able to produce H_2O_2 in the absence of its natural electron acceptor CoQ, but to a lesser extent than MAO [16].

4. α-glycerophosphate dehydrogenase (α-GDH)

α-GDH is located at the outer surface of the inner mitochondrial membrane and catalyzes the oxidation of glycerol-3-phosphate to dihydroxyacetone phosphate, utilizing CoQ as electron acceptor. This reaction is involved in lipid metabolism and in the glycerol phosphate shuttle that regenerates cytosolic NAD^+ from NADH formed in glycolysis. It was found that mitochondria of mouse and Drosophila oxidizing glycerol-3-phosphate produce a measurable amount of H_2O_2 [17].

5. Succinate dehydrogenase (SDH)

SDH is part of complex II and of the citric acid cycle and located in the inner surface of the inner mitochondrial membrane (figures 1.2. and 1.3.) and oxidizes succinate to fumarate using CoQ as electron acceptor. In artificial systems with SDH incorporated in liposomes and in the absence of CoQ, it was shown that SDH can produce ROS [18], but it remains unclear if SDH is able to produce ROS *in vivo* [9].

6. Aconitase (ACO)

ACO is located in the mitochondrial matrix and catalyzes the conversion of citrate to isocitrate as part of the citric acid cycle. Isolated and inactivated ACO, for example by oxidation of its iron-sulfur cluster, induces the production of $^{\bullet}OH$ radicals, most likely mediated by released Fe^{2+} [19].

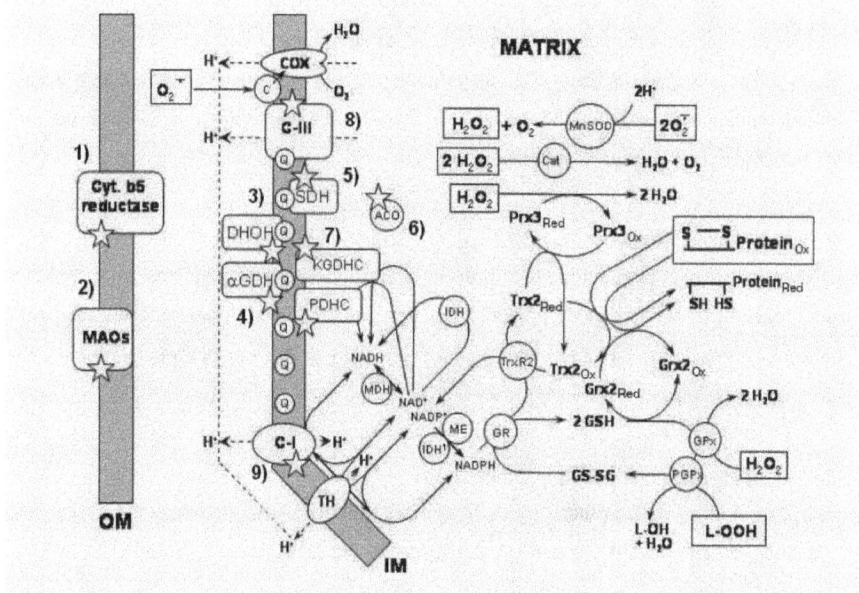

Figure 1.3. Selected ROS-production enzymes and ROS-detoxifying systems inside mitochondria. The nine ROS production sites are marked with their respective number.
(adapted from Andrejev et. al. [9] and slightly modified)

7. α-Ketoglutarate dehydrogenase complex (KGDHC)

KGDHC is, as a part of the citric acid cycle, tightly attached to the matrix side of the inner mitochondrial membrane and catalyzes the oxidation of α-ketoglutarate to succinyl-CoA using NAD^{+} as electron acceptor. The complex consists of multiple copies of three enzymes, α-ketoglutarate dehydrogenase, dihydrolipoamide succinyltransferase and dihydrolipoamide dehydrogenase (DlD). The latter is a flavin-containing enzyme, which was also found to be a part of the pyruvate dehydrogenase complex (PDHC).

- 17 -

Both enzyme complexes, i.e. KGDHC and PDHC, have been shown to generate ROS, namely $^{\bullet}O_2^-$ and $^{\bullet}OH$ radical, in studies using isolated and purified enzymes from bovine heart [20, 21]. It seems very likely that the DlD part of the complex is responsible for the ROS producing properties of both, KGDHC and PDHC. The ROS production by KGDHC is stimulated by low availability of its natural electron acceptor, NAD^+.

8. Complex III of the respiratory chain (C-III, ubiquinone-cytochrome c reductase)

Complex III of the respiratory chain is thought to be the major ROS production site apart from Complex I. Complex III is part of the respiratory chain and serves to oxidize reduced CoQ (ubiquinol, $CoQH_2$ or QH_2), which is carried out via a set of reactions called the "Q-cycle". This process is coupled to a vectorial translocation of protons that serves to maintain $\Delta\Psi$.

It was shown in numerous studies that C-III can produce $^{\bullet}O_2^-$ which is then rapidly dismutated to H_2O_2 by mitochondrial SOD [22-25].

The common model of how $^{\bullet}O_2^-$ is produced is shown in figure 1.4.. $CoQH_2$ is oxidized at the outer leaflet of the inner mitochondrial membrane (Q_o site of C-III) transferring one electron via the Rieske Protein (ISP, iron-sulfur protein) and cytochrome c1 (cyt.c1) to cytochrome c (cyt.c), and onwards to cytochrome c oxidase. The formed unstable semioxidized CoQ (semiquinone, $^{\bullet}Q^-$) donates the second electron via cytochrome b_{low} (b_{low}) and cytochrome b_{high} (b_{high}) to the inner leaflet of the inner mitochondrial membrane (Q_i site of C-III) where it serves to reduce a "new" CoQ molecule. A complete reduction of CoQ, however, requires two electrons. The second electron comes from a second $CoQH_2$ oxidized at Q_o site. The electron originating from the first $CoQH_2$ produces a stable semiquinone moleculeat the Q_i site, which is completely reduced to a "new" $CoQH_2$ by the electron coming from the oxidation of the second $CoQH_2$ at the Q_o site.

It is very likely that the unstable $^{\bullet}Q^-$ at Q_o site is responsible for the ROS production at C-III, in that it may donate its electron to O_2 instead to b_{low}. This theory was substantiated by the finding that addition of antimycin A interrupts the electron transfer to the Q_i site, leading to suppressed reduction of CoQ [26]. This action of antimycin A is thought to enhance the ROS production at C-III due to an accumulation of unstable $^{\bullet}Q^-$ at the Q_o site, favoring the possibility of unwanted electron transfer to O_2 [22, 26, 27].

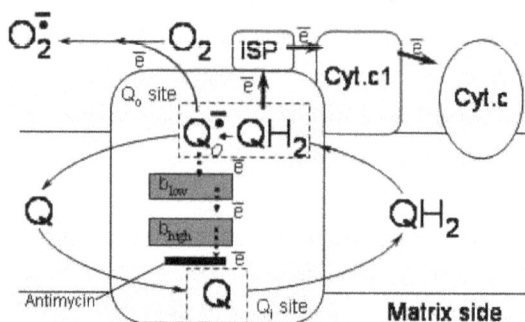

Figure 1.4. Q-cycle model and $^\bullet O_2^-$ formation at C-III.
(*adapted from Andrejev et. al. [9] and slightly modified*)

9. Complex I of the respiratory chain (C-I, NADH-ubiquinone reductase)

Complex I is an integral inner membrane multi-protein complex exposed to both, matrix and intermembrane space. It oxidizes NADH using CoQ as electron acceptor in a reaction coupled with proton pumping and $\Delta\psi$ generation [28].

It has been clearly demonstrated in several studies that C-I can form $^\bullet O_2^-$ in the presence of NADH, but the detailed mechanism remains unclear. Andrejev and co-workers favor the theory that ROS are produced by one of the iron-sulfur centers and not by flavin *per se* [29].

There are three additional major experimental paradigms concerning ROS production attributed to C-I – (i) reverse electron transfer (RET), (ii) rotenone-induced ROS production, (iii) ROS production in normally functioning respiratory chain.

Reverse electron transfer is a process in the respiratory chain that allows electrons to be transferred against the gradient of redox potentials of electron carriers, from CoQH$_2$ to NAD$^+$ instead to C-III and further on to O$_2$. RET produces high amounts of $^\bullet O_2^-$ which are converted to H$_2$O$_2$ and $^\bullet$OH radical, respectively. RET is inhibited by rotenone which blocks reverse electron transfer in close proximity of the coenzyme Q binding site. The whole process of RET is thermodynamically unfavorable and is therefore dependent on the energy obtained from the membrane potential [30]. The reduction of CoQ for this reaction requires FADH$_2$-linked oxidizeable substrates such as succinate or α-glycerophosphate. This energy dependency explains why RET is reliant on high membrane potential and is prevented by the action of uncouplers [31]. A 10% decrease in $\Delta\Psi$ inhibits 90% of ROS production [30, 32, 33]. According to this fact, RET is additionally inhibited by all energy-consuming processes like ATP synthesis [34] or Ca^{2+} uptake [35].

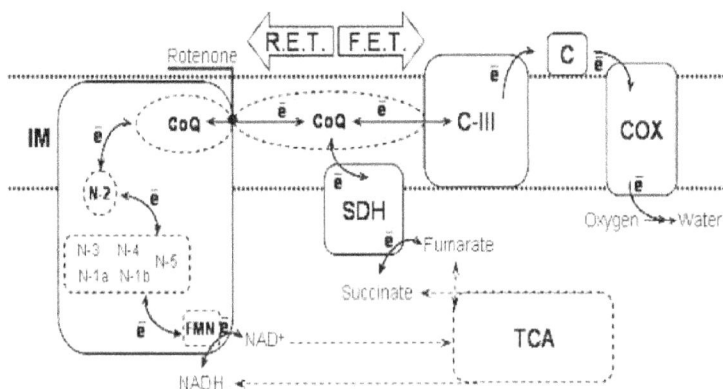

Figure 1.5. Forward and reverse electron transfer in the respiratory chain. The scheme shows the sequence of electron transfer reactions in forward (indicated by arrow F.E.T.) and reverse (R.E.T.) directions. (*adapted from Andrejev et. al. [9] and slightly modified*)

Rotenone-induced ROS production has been studied in intact mitochondria oxidizing NAD^+-linked substrates such as pyruvate and is dependent on a very high degree of reduction of redox carriers upstream of the rotenone binding site. Rotenone-induced ROS production is obviously not regulated by $\Delta\Psi$, as the inhibition of C-I with rotenone ultimately leads to the dissipation of $\Delta\Psi$. Similar effects to rotenone induced ROS production show inhibitors which act downstream of the rotenone binding site (e.g. Complex III inhibitors) because they induce the required highly-reduced state of mitochondrial redox centers and carriers which is necessary for efficient ROS production. It was also found that the redox properties of the rotenone-induced ROS producing site are the same as the N-1a iron-sulfur center of C-I. [36].

ROS production in normally functioning respiratory chain, i.e. in the absence of C-I inhibitors, is supported by oxidation of NAD^+ linked substrates. It is dependent on high $\Delta\Psi$, but to a lesser extent than RET [30, 32, 37].

In all of these three hypothesis ROS generation requires highly reduced states of the involved redox center or redox carriers as well as high $\Delta\Psi$ for *RET induced ROS production* and for *ROS production in normally functioning respiratory chain*.

In conclusion, ROS production at complex I of the mitochondrial electron transport chain is widely accepted as the main source of mitochondrial ROS, but the detailed mechanisms are still unclear and need further investigation [9, 36].

In general the process of unwanted one-electron reduction by electron carriers of the respiratory chain seems to be more likely in the resting-state, when respiration is limited by a lack of ADP [4].

1.3. ROS damage to biological tissues

ROS produced by mitochondria or by any other component inside the cell can damage macromolecules located in close proximity to their origin. To balance their impact, the cell is equipped with an arsenal of ROS-detoxifying systems, which will be discussed in chapter 1.4.. If these systems fail, ROS are no longer detoxified and may react with macromolecules such as lipids, proteins or DNA.

ROS damage to lipids

Lipid membranes, especially the mitochondrial membrane, represent one of the major targets of ROS. The reaction between ROS and unsaturated fatty acids is called lipid peroxidation and is a process carried out in three steps: initiation – propagation – termination [3]. The initiation and propagation process is shown in reaction 1.8. to 1.12., the termination reaction in 1.13..

$$R-CH_2 + {}^\bullet OH \longrightarrow R-CH^{\bullet-} + H_2O$$

Reaction 1.8. Lipid peroxidation - I. During initiation the reactive species (here ${}^\bullet$OH radical) donates its electron to a methyl-group of an unsaturated fatty acid producing a reactive carbonyl radical.

$$R-CH^\bullet + O_2 \longrightarrow R-CHOO^\bullet$$

Reaction 1.9. Lipid peroxidation - II. After molecular stabilization the carbonyl radical can react with O_2 to produce a peroxyl radical.

$$R-CHOO^\bullet + R-CH_2 \longrightarrow R-CHOOH + R-CH^\bullet$$

Reaction 1.10. Lipid peroxidation - III. During propagation the peroxyl radical reacts with other lipid molecules to generate lipid hydroperoxid (R-CHOOH) and carbonly radicals which can in turn react with O_2 to create more peroxyl radicals.

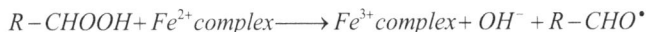

$$R-CHOOH + Fe^{2+}complex \longrightarrow Fe^{3+}complex + OH^- + R-CHO^\bullet$$

Reaction 1.11. Lipid peroxidation - IV. Lipid hydroperoxids are relatively unreactive but in the presence of iron-II complexes it is easily decomposed to form harmful alkoxyl radicals.

$$R - CHOOH + Fe^{3+} complex \longrightarrow R - CHOO^{\bullet} + H^+ + Fe^{2+} complex$$

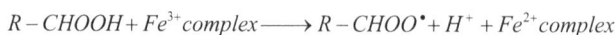

Reaction 1.12. Lipid peroxidation - V. In the presence of iron-III complexes R-CHOOH reacts to form peroxyl radicals, enhancing the propagation process.

$$R - CHOO^{\bullet} + R - CHOO^{\bullet} \longrightarrow unreactive\ product + O_2$$

Reaction 1.13. Lipid peroxidation - VI. The termination reaction is only possible when two peroxyl radicals meet to form an unreactive product and molecular oxygen.

Since lipid peroxidation is a chain reaction, the damage caused by one single ROS molecule can be severe. Oxidation of lipids in mitochondrial membranes contributes to apoptosis induction (see chapter 2.2.) and influences oxidative phosphorylation because it makes the membrane leaky for H^+, which in turn has a critical impact on the maintenance of $\Delta\Psi$ [38].

The cell is equipped with an enzyme called phospholipase A_2, which is able to repair damage caused by lipid peroxidation. Phospholipase A has been shown to remove oxidized lipids from membranes and its activity is enhanced under conditions of oxidative stress [39].

ROS damage to proteins

The damage of ROS to proteins is most severe in mitochondria because mitochondrial proteins are easily oxidized by ROS. For instance superoxide can attack iron sulfur centers in enzymes such as succinate dehydrogenase, NADH ubiquinone reductase and aconitase, thereby destroying their catalytic function and releasing Fe^{2+}, which enhances the formation of $^{\bullet}OH$ radical via the Fenton reaction [40-42].

In addition, ROS may induce serious modifications of various amino acids. Especially amino acids like cysteine or glycine have been shown to be preferentially attacked by ROS, because they can easily oxidize SH-groups (thiol oxidation). This can result in the development of disulfide bridges (S-S bounds) leading to modifications of the protein structure. This process seems to be relevant for the onset of mitochondrial permeability transition and apoptosis induction, which will be discussed in chapter 2.2..

ROS damage to DNA

Oxidative damage to DNA covers single and double strand breaks, molecular modifications of the bases and of the deoxyribose backbone as well as cross-linkage of DNA to other molecules, e.g. to proteins [3]. Damage of mitochondrial ROS to nuclear DNA seems quite unlikely because the nucleus may be out of the radius of action of ROS originating from mitochondria. Indeed, nuclear DNA is only affected in the case of immense ROS production in mitochondria [3]. Furthermore the cell possesses various repair mechanisms to avoid the replication of permanently modified DNA. However, if ROS damage to nuclear DNA is irreversible, the outcome is critical, because DNA modifications are potentially mutagenic, contributing to cancer development.

In contrast, nucleic acids of mitochondrial DNA (mtDNA) are much more exposed to mitochondrial derived ROS. The level of oxidatively modified bases in mtDNA is estimated to be about 20-fold higher than in nuclear DNA [43]. There are many indications that oxidative damage of mtDNA plays a crucial role in some mitochondrial diseases, aging and cancer [5]. It is likely that respiratory dysfunctions induced by mitochondrial genomic instability are caused by mitochondria-derived ROS. Mitochondrial DNA encodes for thirteen polypeptides, twenty two transfer RNAs (tRNAs) and two ribosomal RNAs (rRNAs), all of which involved in the composition of the oxidative phosphorylation machinery. To build up a functional respiratory chain, nuclear encoded proteins and mtDNA encoded proteins have to be assembled. Logically, ROS-modified mtDNA can encode for malfunctioning proteins, eventually leading to cell injury through the disruption of electron transport, $\Delta\Psi$ and ATP generation.

1.4. Cellular ROS-defense mechanisms

As described above, ROS can be generated as a result of normal metabolic processes after induction by internal or external stimuli. Under normal conditions, antioxidant systems minimize the damaging effect of ROS. However, if the intracellular ROS concentration increases drastically and can no longer be counterbalanced the antioxidant defense mechanisms are overpowered and massive oxidative stress occurs.

In general, the cell can protect itself against oxidative damage by using the detoxifying abilities of antioxidants and mechanism which reduce the probability of ROS-overproduction.

Common antioxidant systems

Manganese-containing superoxide dismutase (MnSOD)

MnSOD is located in the mitochondrial matrix where it dismutates $^{\bullet}O_2^-$ to H_2O_2 (reaction 1.5.) [44]. Unless H_2O_2 reacts with Fe^{2+} to form harmful $^{\bullet}OH$ radicals it might, due to its membrane permeable qualities and relatively long lifetime, escape from mitochondria without causing any immediate damage. Thus, the main function of MnSOD is to transform $^{\bullet}O_2^-$ to H_2O_2 to generate an ROS type that is more likely to be released from mitochondria than to cause direct damage to sensitive parts of the respiratory chain, e.g. iron sulfur cluster containing enzymes.

MnSOD does not require any cofactors and the efficiency of its action is only limited by its amount. It was shown *in vivo* that transgenic mice lacking MnSOD do not survive longer than a few days, whereas animals with only partial MnSOD activity are viable and fertile, but were more susceptible to cancer development [45, 46].

Apart from MnSOD three other classes of SOD have been identified to date - Cu/Zn-SOD, Ni-SOD and extracellular SOD - all of which catalyze the same reaction, but in different cellular compartments [8].

Catalase (Cat)

Catalase is an enzyme located in mitochondria and in peroxisomes and converts H_2O_2 to O_2 and H_2O [47]. Thus, catalase supports MnSOD in order to avoid the presence of ROS inside the mitochondria because it is able to detoxify significant amounts of H_2O_2 even before its efflux into the cytosol.

$$H_2O_2 + H_2O_2 \xrightarrow{\;Cat\;} 2H_2O + O_2$$

Reaction 1.14. Conversation of H_2O_2 to O_2 and H_2O by catalase activity.

Catalase is one of the most efficient enzymes known and can not be saturated by H_2O_2 at any concentration [48]. Cellular overexpression of catalase has been shown to protect against oxidative injury [49] and induce apoptosis-resistance in thymocytes [50].

Transgenic mice lacking catalase appear normal but are slightly more sensitive to oxidative stress [51]. In yet another study H_2O_2 removal by catalase was found to be irrelevant compared to the action of other H_2O_2 detoxifying enzymes [52].

However, the detailed aspects of the biological significance of catalase in the ROS-defense network is still not completely understood.

Cytochrome c (cyt.c)

Cytochrome c is a multifunctional enzyme acting as electron carrier in the respiratory chain and plays a major role in mitochondrial apoptosis. In addition it is known to regenerate molecular oxygen from superoxide [4, 38]. Cyt.c is located in the intermembrane space of mitochondria. Only a small part of the cyt.c molecules, i.e. those which are membrane-bound, contribute to the electron transfer of the respiratory chain, the unbound molecules are sampling the intermembrane space for $^\bullet O_2^-$ to be detoxified.

$$cyt.c_{oxi} + {}^\bullet O_2^- \longrightarrow cyt.c_{red} + O_2$$

Reaction 1.15. Regeneration of O_2 from $^\bullet O_2^-$ by cyt.c.

Reduced cyt.c (cyt.c_{red}) can be "regenerated" to oxidized cyt.c (cyt.c_{ox}) following the reaction below.

$$4\,cyt.c_{red} + O_2 + 4H^+ \xrightarrow{\;COX\;} 4\,cyt.c_{ox} + 2H_2O$$

Reaction 1.16. Reoxidation of reduced cyt.c by O_2 via the action of cyt.c oxidase.

Following the hypothesis of Skulachev and his coworkers cyt.c action represents the most effective means of scavenging $^\bullet O_2^-$ because $^\bullet O_2^-$ formed from O_2 is converted back to O_2. In addition cyt.c can be reoxidized as shown above and is then ready to resume its detoxifying action. This oxidation of cyt.c by cytochrome c oxidase (COX) contributes to the generation of $\Delta\Psi$, representing a intriguing ability of this antioxidant system to generate useful metabolic energy while detoxifying harmful superoxide [4, 53].

Coenzyme Q

Coenzyme Q is a lipid soluble, vitamin-like substance commonly known as electron carrier in the respiratory chain. It can exist in three forms inside the cell - as reduced coenzyme Q (ubiquinol, $CoQH_2$ or QH_2), as semi-reduced or semi-oxidized coenzyme Q (semiquinone, $^\bullet Q^-$), or as oxidized Coenzyme Q (ubiquinone, CoQ or Q). Apart from being critically involved in the electron transport inside mitochondria, ubiquinol represents an important antioxidant, acting close to mitochondria and lipid membranes, where it serves to neutralize lipid peroxyl radicals by donating one of its hydrogen atoms to become semiquinone, which is then restored to ubiquinol by the Q-cycle in C-III of the respiratory chain.

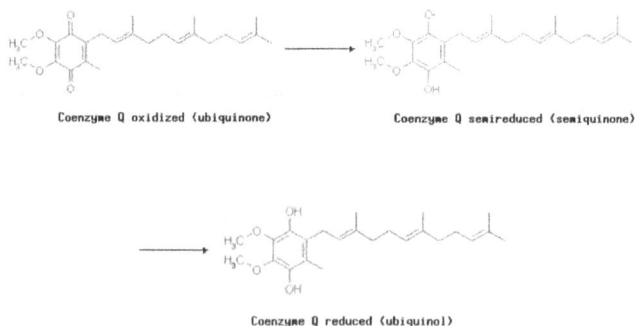

Coenzyme Q oxidized (ubiquinone)   Coenzyme Q semireduced (semiquinone)

Coenzyme Q reduced (ubiquinol)

Figure 1.6. Molecular structure of coenzyme Q in its three different forms.

Another important function of ubiquinol is its ability to regenerate α-tocopherol from its oxidized form. Ubiquinol has been shown to inhibit protein and lipid oxidation in cell membranes and to minimize DNA modifications [54]. It has been successfully used in cancer prevention and therapy [55], to control blood pressure in humans [56] and for treatment of mitochondrial neuromuscular diseases [57, 58]. Moreover, a protective effect of ubiquinol in neurogenerative deseases like Parkinson´s disease and dementia has been demonstrated [59].

Ascorbic acid

Ascorbic acid (vitamin C) is a water-soluble molecule found in various nutrient sources e.g. vegetables and fruits, representing one of the most important exogenous antioxidants. Ascorbic acid functions as ROS-quencher. It can be directly oxidized by ROS, which are in turn detoxified.

Principally, ascorbic acid can react with all kinds of ROS but seems to have a special affinity to scavenge ROO$^\bullet$ [60]. In addition, ascorbic acid is thought to protect against lipid peroxidation by regenerating reduced α-tocopherol (from its oxidized form) [61]. Once ascorbic acid is oxidized to dehydroascorbic acid it can be converted back to its reduced form by reduced glutathione (GSH) and NADPH.

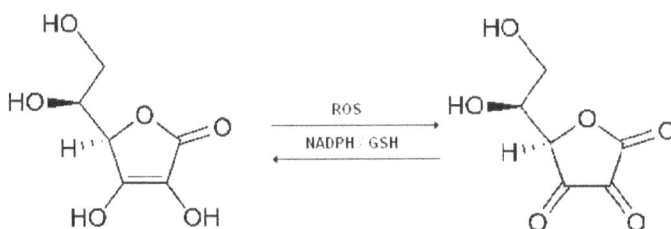

Figure 1.7. Molecular structure of ascorbic acid and dehydroascorbic acid.

Vitamin C is used as dietary supplement because it is thought to reduce the risk and slow down the development of certain diseases, including cancer, age-related eye diseases as well as coronary heart diseases [60].

α-tocopherol

Beside ascorbic acid α-tocopherol, an isoform of vitamin E, is the most important vitamin with antioxidant properties. As a lipid-soluble, free radical scavenging antioxidant α-tocopherol is mainly present in mitochondrial membranes. It acts to remove lipid peroxides by reducing peroxyl- and alkoxyl radicals in order to stop the lipid peroxidation process. The oxidized form of α-tocopherol is regenerated to the reduced form by ubiquinol (in membranes) and ascorbic acid (at the membrane-water interface).

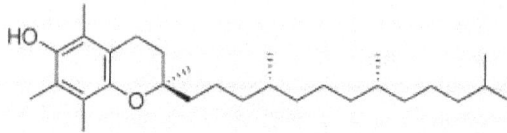

Figure 1.8. Molecular structure of α-tocopherol.

Like vitamin C, vitamin E is widely used as a supplemental antioxidant for co-treatment of various diseases. It is often applied in combination with vitamin C and glutathione because these three major antioxidants are linked in a kind of "antioxidant-network" (figure 1.9.) [62, 63].

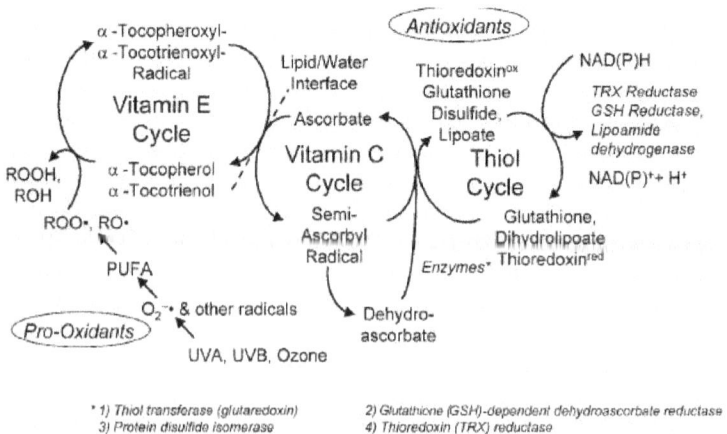

Figure 1.9. The antioxidant network involving vitamin E, vitamin C and thiol redox cycles.
(adapted from Packer et. al. [63])

Carotenoids

Carotenoids are molecules synthesized inside various vegetables and fruits in which they are responsible for the bright colors. Most known carotenoids have antioxidant properties. Especially β-carotene has been shown to antagonize oxidative stress in various harmful situations.

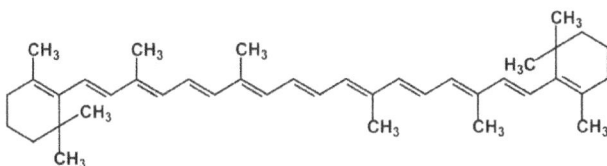

Figure 1.10. Molecular structure of β-carotene.

Recent studies have demonstrated antioxidant activity of β-carotene in lipid membranes, but it was shown to be less effective in ROS detoxification than α-tocopherol [64].
The antioxidant properties of β-carotene and other carotenoids may vary depending on the system of use. A mixture of carotenoids with other antioxidants may increase their ability to protect against lipid peroxidation [65].

Bcl-2

The proto-oncogene Bcl-2, which was first found in a form of breast cancer lymphoma is commonly known as anti-apoptotic member of the Bcl-2 protein family, consisting of 20 members with either pro- or anti-apoptotic features. In numerous studies Bcl-2 has been shown to be able to attenuate or even abolish programmed cell death and the majority of studies focused on its interaction with and regulation of the mitochondrial permeability transition pores (mPTPs) [66-68].
Bcl-2 might also possess antioxidant properties. Two independent studies have shown that altered Bcl-2 expression influences the cellular GSH pool [69, 70]. These studies indicate that Bcl-2 may modulate apoptosis by interfering with GSH-biosynthesis.
Other studies have demonstrated that Bcl-2-overexpression may protect cells from lipid peroxidation and thiol oxidation induced by menadione and hydrogen peroxide, respectively [71, 72], but the role of Bcl-2 in the antioxidant network requires further investigations.

NAD(P)H

Skulachev has hypothesized that the decomposition of $NAD(P)^+$ may be, under special circumstances, a way to compensate for ROS overproduction [73]. The ROS-activated enzyme NAD^+ glycohydrolase is localized in the outer mitochondrial membrane. In case of short-term mPTP opening, this enzyme catalyzes the decomposition of released NAD^+ and $NADP^+$. The following exhaustion of the $NAD(P)^+$ pool must, according to Skulachev, stop mitochondrial ROS-production [73]. Other studies claim that NAD(P)H *per se* can serve as non-enzymatic antioxidant, but this theory has not been definitely proven yet [74].

Glutathione

Glutathione (GSH) is a tripeptide (L-γ-glutamyl-L-cysteinyl-glycine) distributed ubiquitinously among animals, plants and microorganisms and represents the probably most efficient antioxidant in higher organisms [75]. GSH levels in human cells range from approximately 0.1 mM up to 10 mM [75]. GSH is a hydrophilic molecule that is distributed virtually everywhere in the cell, with 10-12% of its total content found in mitochondria. Due to the relatively small mitochondrial matrix volume the concentration of GSH in mitochondria is higher than in the cytoplasm, ranging from 2 to 14 mM [75].

GSH-biosynthesis takes place in the cytoplasm. Certain transport systems like GSH-transporters or dicarboxylate- and 2-oxoglutarate-carriers are responsible for GSH-flux into mitochondria [9]. Due to its high intracellular concentration and its negative redox potential, GSH generates an immense reducing power. The thiol-group (-SH) in its molecular structure can be easily oxidized by practically all free radicals, e.g. $^•OH$, $RO^•$, $ROO^•$ and other oxygen-centered radicals.

$$2\,GSH + 2\,R^• \longrightarrow 2\,RH + GSSG$$

Reaction 1.17. Formation of GSSG after the reaction of GSH with a radical-molecule.

After oxidation by a radical ($R^•$), two oxidized GSH molecules react to a stable form of oxidized glutathione, named GSSG. GSSG can be regenerated to GSH by the mitochondrial enzyme glutathione reductase (GR), using NADPH as a source of reducing equivalents [76].

$$GSSG + 2\,NADPH \xrightarrow{\ GR\ } 2\,GSH + 2\,NADP^+$$

Reaction 1.18. Regeneration of GSH by glutathione reductase.

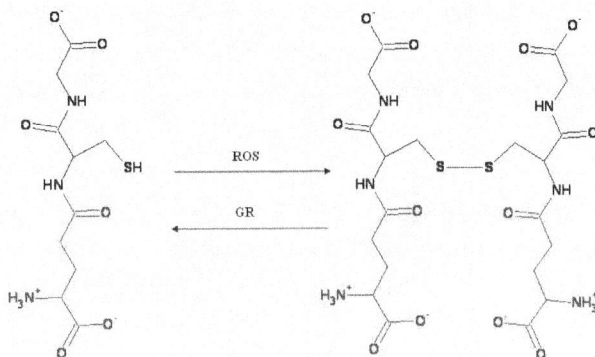

Figure 1.11. Molecular structure of GSH and GSSG.

In addition to its property to detoxify free radicals, GSH is known to effectively decompose H_2O_2, a reaction catalyzed by the enzyme glutathione peroxidase (GP).

$$2GSH + H_2O_2 \xrightarrow{\quad GP \quad} 2H_2O + GSSG$$

Reaction 1.20. Decomposition of H_2O_2 by GSH via glutathione peroxidase.

The protein family of hydrogen peroxidases uses GSH as source of reducing equivalent. The propbabely most important representative of this protein family is phopholipid hydroperoxide glutathione peroxidase (PGP), a selenoenzyme which is selective to phospholipid hydroperoxides, H_2O_2 and cholesterol peroxides [77, 78]. PGP is the only enzyme of the GP-family which is known to reduce peroxidized phospholipids inside membranes and is therefore thought to play an important role in the overall ROS-detoxifying network [79].

$$2GSH + R - CHOOH \xrightarrow{\quad PGP \quad} 2R - CHOH + GSSG$$

Reaction 1.21. Decomposition of lipid ROOH by GSH via phospholipids hydroperoxide glutathione reductase.

The mitochondrial enzyme glutathione-S-transferase (GST) uses GSH as a co-factor in conjugation reactions [80]. GST conjugates GSH to endogenous substances (e.g. estrogens or toxic products of lipid peroxidation), exogenous electrophiles (e.g. organic halides) and various xenobiotics, in order to protect mitochondria from several types of toxins. The initial products of such detoxification reactions are stable sulfides of GSH. In following set of reactions, L-glutamate and glycine residues are removed, forming S-substituted L-cysteines. After acetylation the cysteinyl group can form mercapturic acid, which can then be easily excreted with urine [81].

With regard to the facts presented above, the maintenance of a high intracellular concentration of GSH seems to be indispensable for the survival of the cell. The availability of GSH in nutrition is limited (less than 150 mg/day) [82], meaning that the intracellular mechanisms of GSH generation (the regeneration of GSH by glutathione reductase and GSH *de novo* synthesis) must not be impaired.

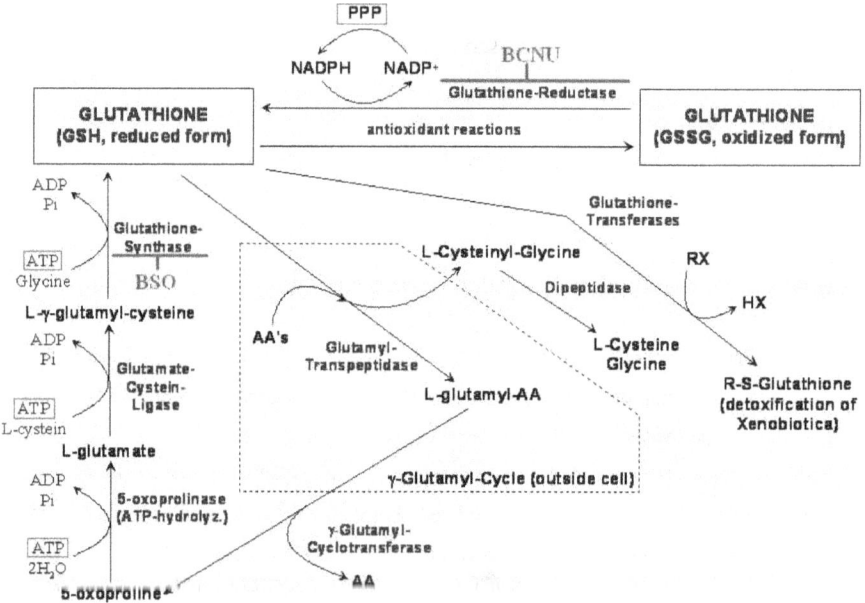

Figure 1.12. GSH biosynthesis.

Figure 1.12. shows the biosynthesis of GSH. GSH regeneration by glutathione reductase is connected to the pentose phosphate pathway (PPP) which provides NADPH to reduce GSSG.

In GSH *de novo* synthesis there are three enzyme-catalyzed reactions which are highly dependent on the availability of ATP. These are (i) the reaction of 5-oxoproline to L-glutamate catalyzed by 5-oxoprolinase, (ii) the ligation of the L-glutamate and cystein residue via glutamate-cystein ligase and (iii) the addition of glycine by glutathione synthetase. This ATP-dependency and the fact that the metabolization of glucose in the PPP represent the most important source of NADPH provide evidence that GSH-biosynthesis is highly dependent on the energetic status of the cells.

There are a number of chemical agents which may interfere with GSH biosynthesis. Among them DL-buthionine-sulfoximine (BSO) and 1,3-bis-(2-chloroethyl)-1-nitrosourea (BCNU). BSO is an irreversible inhibitor of glutathione synthetase [83] and BCNU has been demonstrated to inhibit glutathione reductase [84] (figure 1.12.).

A change of the intracellular GSH level has several consequences for the cell. Primarily, GSH depletion has been reported to increase intracellular ROS levels, followed by apoptosis induction [85-87]. In addition, depletion of the intracellular GSH pool may also render cells more susceptible to apoptosis induced by various stimuli [87, 88]. In contrast, increased intracellular GSH levels have been shown to protect cells against ROS overproduction and subsequent cell death [86, 89, 90].

GSH is an essential mediator of several human diseases, including cancer and cardio-vascular diseases [91, 92]. Manipulation of the intracellular GSH level has been successfully used in both cancer prevention and cancer therapy [92, 93].

In comparison to all other antioxidants discussed in this chapter, GSH, its metabolism and all GSH-utilizing enzymes appear to be the leading actors in the play of ROS-detoxifying mechanisms [91, 94].

Artificial antioxidants

In the last years researchers have developed a variety of different artificial substances with the intention to mimic features of antioxidant enzymes or to simply provide means for direct de-radicalization. Some of these artificial antioxidants, which were used for this doctoral thesis will be discussed briefly.

In one of the individual studies of this doctoral thesis, the role of ROS in apoptosis induced by Ro-31-8220, an analog of staurosporine, was investigated (chapter VI). We employed four artificial antioxidants to analyse their effect onRo-31-8220 mediated cell death.

Two of these substances, MnTBAP and Euk-8, have been designed to imitate antioxidant enzymes. MnTBAP is a cell permeable metalloporphyrin which mimics mitochondrial SOD and, in addition, has potent hydrogen peroxide scavenger properties [95-97]. Likewise, Euk-8, a synthetic selen-manganese complex, has both SOD and catalase activities [98, 99].

In addition, two other artificial antioxidants have been used, both known for their ability to efficiently attenuate ROS overproduction and concomitant apoptosis. Trolox is a water-soluble analog of α-tocopherol with potent ROS quenching properties, especially of hydrogen peroxide. Trolox has been shown to support protein stability by inhibiting lipid peroxidation [100, 101]. Glutathione monoethyl ester (GSH-ee) is intracellularily converted to GSH and has been successfully used in various applications to increase intracellular GSH levels [92, 102].

Cellular mechanisms to avoid ROS-overproduction

The cellular strategies described below are usually only executed in the case of excessive ROS production, where the conventional antioxidant mechanisms fail.

There are three major strategies for the cell to prevent excessive ROS generation by mitochondria: (i) down-regulating the concentration of intracellular molecular oxygen in order to avoid unspecific one-electron reduction, (ii) aconitase-mediated control of hydrogen supply to the respiratory chain and (iii) mild-uncoupling.

The control of molecular oxygen concentration

The concentration of O_2 inside the cell determines the probability of intracellular generation of $^\bullet O_2^-$. The higher the concentration of O_2, the higher is the chance of unspecific one-electron reduction. Vertebrates have developed several strategies to control the intracellular concentration of O_2 and, therefore, the possibility of $^\bullet O_2^-$ production (reviewed by Papa and Skulachev [5, 103]).

Aconitase-mediated control of hydrogen supply to the respiratory chain

All reduction processes of O_2 inside mitochondria, including the production of $^\bullet O_2^-$, are deemed to stop when the hydrogen supply to the respiratory chain is eliminated. According to Skulachev, this strategy is a very radical one, and is thought to become active in situations characterized by enormous ROS production [4]. One of the possible mechanisms to stop hydrogen supply might be the reversible inactivation of aconitase by $^\bullet O_2^-$. $^\bullet O_2^-$ oxidizes one of the four Fe^{2+} ions in the iron-sulfur cluster present in active aconitase. As aconitase initiates the Krebs cycle, its oxidation leads to an inhibition of the Krebs cycle at its very beginning. Therefore, the main dehydrogenase reactions that supply the respiratory chain with reducing equivalents can not be carried out. This entails a very strong oxidation of respiratory chain electron carriers capable of one-electron O_2 reduction, decreasing the possibility of $^\bullet O_2^-$ production [4].

Mild-uncoupling

The most effective strategy to avoid unwanted ROS production inside mitochondria is the concept of "mild-uncoupling" [4, 38, 104-106].

As described in chapter 1.2., ROS are thought to be produced most efficiently in the resting state of mitochondria, when the main mechanism of O_2 consumption (i.e. phosphorylating respiration) is switched off due to the exhaustion of ADP and the accumulation of ATP. Concomitant with ADP depletion, intracellular O_2 concentration increases because of the inhibition of respiration. At the same time the lifetime of $^\bullet Q^-$ becomes much longer due to an inhibition of the Q-cycle, therefore increasing the probability of one-electron reduction.

The solution of this problem is partial uncoupling (mild-uncoupling), decreasing $\Delta\Psi$ and stimulating O_2 consumption, which in turn lowers concentration of O_2, shortens $^\bullet Q^-$-lifetime and eventually inhibits ROS production. Logically, mild-uncoupling followed by a decrease in $\Delta\Psi$ makes the production of $^\bullet O_2^-$ less likely because the major $^\bullet O_2^-$-generating-sites, e.g. complex I and complex III of the respiratory chain depend on a high $\Delta\Psi$.

Skulachev and his group have demonstrated that a 15% decrease of $\Delta\Psi$ caused by an uncoupler results in a 10-fold decrease of H_2O_2-production of heart-cell mitochondria [30]. It was also shown that free fatty acids are good candidates for the role of mild uncouplers, because at low concentrations they may operate as compounds able to slightly increase H^+ conductance of the inner mitochondrial membrane and to keep $\Delta\Psi$ below a certain threshold required for efficient ROS production [31].

2. Involvement of ROS in apoptosis induction

2.1 Biochemistry of apoptosis

Apoptosis is a form of non-traumatic cell death, which - as apposed to necrosis - does not result in tissue inflammation. Apoptosis is characterized by a series of energy-dependent biochemical events which lead to typical cell shrinkage and, eventually, to the formation of so-called *apoptotic bodies*. These small vesicles are packed with intracellular components and degraded DNA, ready to be disposed by the immune system, specifically by phagocytes.

Apoptosis is associated with pathologic cell damage, but it also represents an evolutionary conserved process to get rid of superfluous cells during ontogenesis. In this context, a well known example for apoptosis is the differentiation of fingers and toes in a developing mammalian embryo. In an average human body, 50 to 70 billion cells die by apoptosis every day, and the mass of proliferation and subsequent destruction of cells per year is equal to an individuals body weight [107].

Another very important function of apoptosis is to remove cells that might be harmful for the organism, e.g. cells with metabolic malfunctions or cells with critical levels of DNA damage that bear the potential of malignancies.

Caspases – the key molecules in apoptosis

The leading actors in the apoptotic scenario are special types of proteases called caspases (*c*ysteinyl-*asp*artate-specific-prote*ases*), which are mainly responsible for the initiation and execution of the apoptotic response and, eventually, for the decomposition of the intracellular components. Caspases are a family of proteases with cystein in their active center and a specificity for aspartate residues. They are highly conserved throughout evolution and can be found in organisms ranging from insects to humans [68, 107-109].

Fourteen different caspase family members have been identified so far; not all of them are involved in apoptosis. For example, caspase-1 (interleukin-1β (IL-1β) converting enzyme / ICE) has been shown to represent a mediator of inflammatory responses because it can process inactive IL-1β to its active form, which is necessary for its secretion by phagocytes at the site of infection [110].

Most caspases, however, are tightly linked to apoptosis. Caspases are constitutively present in the cell in their inactive "procaspase" form. Upon apoptotic stimuli, procaspases are activated by proteolytic cleavage, which may be executed either by autocleavage or by other caspases.

Caspases which are usually activated through protein-protein interaction and subsequent autocleavage are called initiator caspases (e.g. caspase-8 and -9) whereas caspases activated proteolytically by an upstream caspase are referred to as effector caspases (e.g. caspase-3, caspase-6 and -7) [111]. The consecutive activation of other caspases during programmed cell death creates a growing cascade of apoptotic proteases, leading to decomposition of the cell content.

Caspases have been shown to be responsible for the observed morphological changes during apoptosis [68, 112]. Quite recently, caspases have been found to be also responsible for DNA-laddering, i.e. controlled degradation of DNA by a specific DNase. It was shown, that this DNA ladder nuclease pre-exists in living cells as an inactive complex with an inhibitory subunit. Caspase-3 is able to cleave of this inhibitory subunit, resulting in a release of the activated catalytic subunit. The activated DNA ladder nuclease is called caspase-activated DNase (CAD) whereas the inactive form is named ICAD (inhibited caspase –activated DNase) [113, 114].

Two common models of apoptosis induction

In principal, apoptosis induction can be mediated by either extracellular or intracellular stimuli. In this context, two major apoptotic pathways are known: the *death receptor pathway* and the *mitochondrial / stress-induced pathway* (see figure 2.3) [68].

The death receptor pathway

The common model of this pathway is characterized by the binding of members of the tumor necrosis factor (TNF) family to the so-called death receptor, thus initiating a clustering of the receptor and the formation of a death-inducing signaling complex (DISC) including the adaptor protein FADD (Fas-associated death domain protein). FADD recruits many procaspase-8 molecules that bind to the complex. This close proximity of the procaspase-8 molecules initiates their activation by autocleavage, resulting in the generation of active caspase-8 molecules. Depending on the cell type the signal propagation may follow two routes [115]:

Type I cells are characterized by robust caspase-8 activation, which subsequently leads to the direct activation of other downstream caspases, in particular caspase-3, which is known as the most important caspase in the whole apoptosis program.

Type II cells, in contrast, produce lower levels of caspase-8 which are hardly sufficient to directly activate other caspases. To overcome this handicap, caspase-8 may cleave Bid, a cytosolic member of the Bcl-2 family of apoptotis regulators, to its activated form tBid (truncated Bid) which is in turn able to act on mitochondria and stimulate the mitochondrial apoptosis pathway. Thus, both apoptosis pathways are closely interrelated by the dual action of caspase-8 – a phenomenon termed "cross talk" (figure 2.3) [112, 116, 117].

The mitochondrial pathway

The mitochondria-mediated apoptosis pathway is activated in situations of intense stress, and is therefore also called mitochondrial stress pathway (see figure 2.3.). The stress stimuli can either be external (e.g. chemotherapeutic drugs like etoposide), or internal (e.g. ROS overproduction, severe DNA-damage). These stimuli trigger the first step in mitochondrial apoptosis, a permanent opening of the mPTPs followed by the subsequent collapse of $\Delta\psi$ [108].

The opened mPTPs are permeable for molecules up to 1,5 kDa. As a result, the osmotic balance across the mitochondrial membrane disappears and water from the intermembrane space penetrates into the matrix. As a consequence, the inner mitochondrial membrane expands and due to its large area it ruptures the outer membrane. Different proteins located in the intermembrane space of mitochondria are then released into the cytosol and function as initiators of the apoptotic program [4, 105].

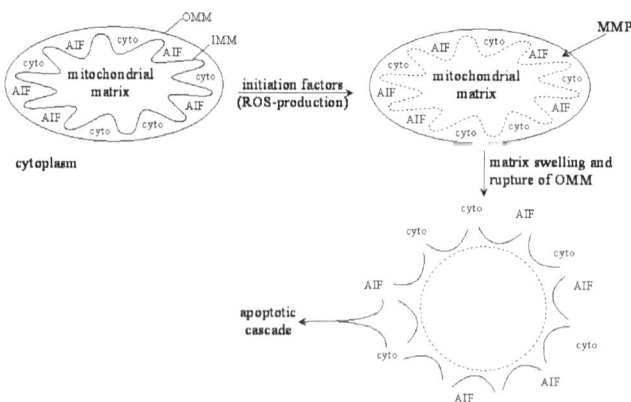

Figure 2.1. Stress-induced opening of mPTPs and the following release of intermembrane proteins.

Cytochrome c is maybe the most important protein released from mitochondria. Cyt.c was already described earlier in this work as electron carrier of the respiratory chain and as antioxidant molecule. Moreover, cyt.c release is regarded as an almost universal feature of apoptotic cell death, although apoptosis induced by death receptors sometimes bypasses the mitochondrial pathway, with no detectable cyt.c release [118]. Cytosolic cyt.c forms a complex, the so called apoptosome, together with the cytosolic protein apoptotic protease activating factor (Apaf) -1, dATP and procaspase-9. The apoptosome functions to activate procaspase-9 by autocleaving processes. Caspase-9 then activates caspase-3 and other down-stream effector-caspases. Caspase-3 activation is where the two apoptotic pathways converge and it is caspase-3 which seems to be mainly responsible to set the apoptotic cascade running [119].

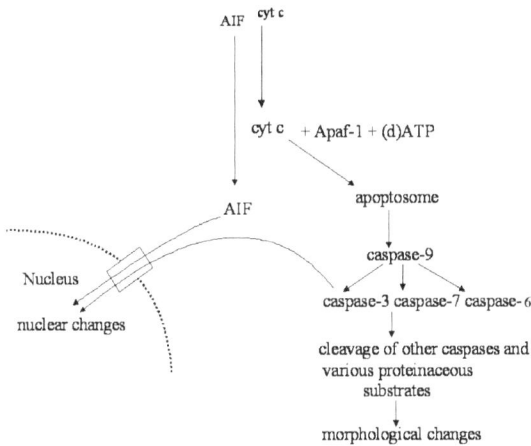

Figure 2.2. Intermembrane proteins initiate the apoptotic cascade.

Another protein which is released from mitochondria is the flavoprotein *apoptosis inducing factor (AIF)*. In viable cells AIF is found exclusively in mitochondria. After it is released to the cytosol it is thought to act on nuclear DNA. It may trigger chromatin condensation and induce cleavage of DNA into high molecular weight fragments [120].

Endonuclease G is a nuclear encoded mitochondrial protein which is released from mitochondrial intermembrane space concomitantly with cyt.c and AIF. In intact mitochondria it seems to be involved in the mitochondrial DNA replication. Like AIF, endonuclease G is thought to contribute to DNA fragmentation [121, 122].

Another protein released from mitochondria during the initial phase of apoptosis is *Smac* (for second mitochondrial-derived activator of caspases)/*DIABLO* (direct IAP-binding protein with low pI). Upon entering the cytosol, Smac antagonizes the action of IAP (*i*nhibitor of *a*poptosis *p*roteins) by attaching itself to them and suppressing their apoptosis-inhibitory function [112, 116].

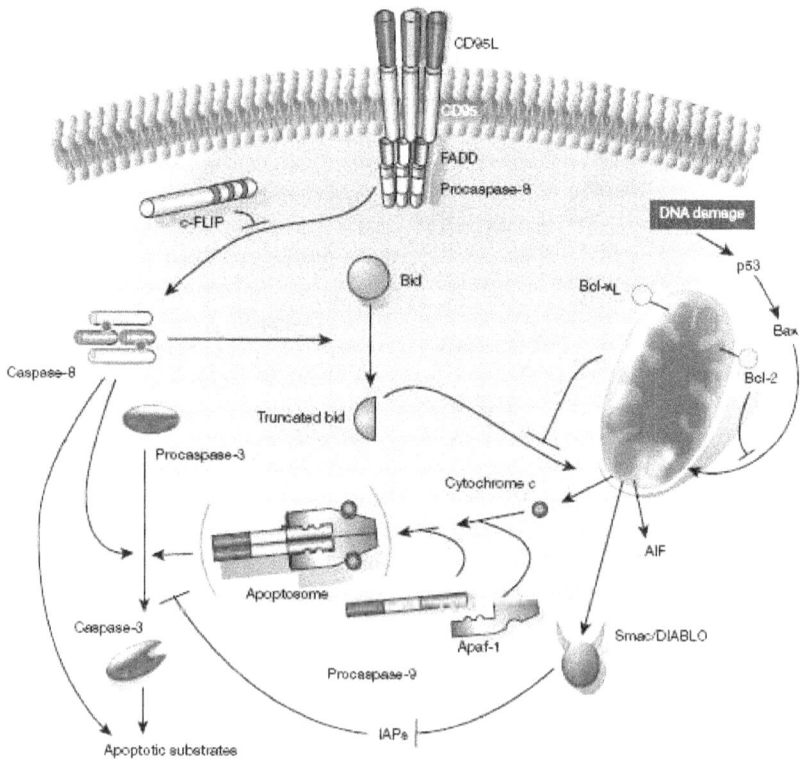

Figure 2.3. Receptor-mediated and mitochondrial apoptosis pathway.
(adapted from Hengartner et. al. [112])

2.2 ROS–mediated induction of mitochondrial apoptosis

The involvement of ROS in apoptosis is manifold. There is the opinion that in most forms of cell death ROS represent essential intermediate messenger molecules [123-125]. Various apoptotic stimuli (such as TNFα, lipopolysaccharide (LPS) or ceramide) have been demonstrated to involve intracellularily generated ROS in their apoptosis inducing action [126, 127]. Even cell death following severe DNA damage was shown to be mediated by apoptosis characterized by ROS overproduction. A critical relationship between the oncogen p53, mitochondrial generated ROS and the induction of apoptosis has been demonstrated recently [128]. Furthermore, there is evidence that caspase-3 is able to induce ROS production which can be regarded as a kind of a positive feedback loop to enhance the apoptotic cascade [129].

However, the overall involvement of ROS in apoptosis is still very complex, but the mechanism of ROS-mediated apoptosis induction seems to be evolutionary conserved and highly related to the opening of the mPTPs [38, 105].

The mitochondrial permeability transition pore (mPTP) complex

The mPTP is a multiprotein complex spanning the inner and outer mitochondrial membrane. In vital cells the complex does not show any pore properties. Upon various apoptotic stimuli the pore changes its molecular structure and forms a pore which is permeable for molecules up to 1.5 kDa. The complex consists of the voltage dependent anion channel (VDAC), the adenine nucleotide translocator (ANT), the benzodiazepine peripheral receptor (BPR) and cycophilin D (Cph.D).

The ANT is located in the inner mitochondrial membrane and normally functions as an antiporter to transport ADP into and ATP out of the matrix [130, 131].

The VDAC is located in the outer membrane. The general function of the VDAC is to permit low-MW molecules to access the solute-specific transport system of the inner membrane. Moreover, VDAC is thought to cooperate with the ANT in mitochondrial nucleotide translocation [131].

The BPR is located in the outer membrane too and controls the transfer of extramitochondrial cholesterol to the inner membrane [132]. Some studies have demonstrated that the binding of different ligands to the BPR facilitates pore opening and apoptosis induction [133] whereas other studies propose an anti-apoptotic effect caused by ligand-receptor binding events [134].

Cph.D is located in the matrix was characterized as a mitochondrial peptidyl prolyl-*cis, trans*-isomerase with a putative role in protein folding, but its detailed function in vital cells has not yet been discovered yet [135]. In apoptosis Cph.D seems to be involved in Ca^{2+}-induced permeability transition [130].

Several other proteins have been found to be associated with the complex, for example creatine kinase (CK) and hexokinase (HK), but their functions in apoptosis induction are largely unclear [130, 131].

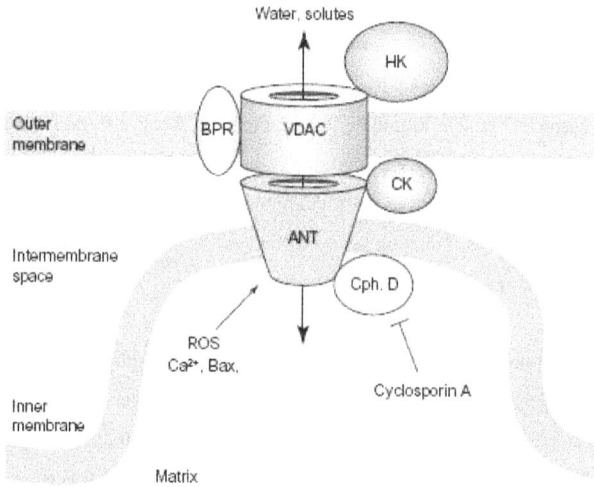

Figure 2.4. The mitochondrial permeability transition pore.
(adapted from Desagher et .al. [135])

Induction of mPTP opening

The protein complex can be converted to a high-conductance unselective channel as response to different factors, i.e. mitochondrial Ca^{2+} overload, reduced concentrations of adenine nucleotides, inorganic phosphate (P_i), proteins of the Bcl-2 family, low $\Delta\psi$ and ROS. In contrast specific inhibitors such as bongkrekic acid (an antagonist of ANT) or cyclosporine A (a ligand of Cph.D) have been demonstrated to efficiently prevent pore opening, membrane depolarization and apoptosis induction in various studies [136, 137].

Ca^{2+} *overload* in the mitochondrial matrix is a major feature of cellular stress and is known to be required in nearly all situations of pore opening [130]. Ca^{2+} is assumed to induce pore opening by binding to low-affinity sites of the complex [138] and by disturbing the integrity of the membrane's lipid layer [139]. The intracellular level of free Ca^{2+} is normally kept low and mitochondrial Ca^{2+} overload is generally unlikely. However, upon direct damage to the ER, which is the cell's largest Ca^{2+} reservoir, Ca^{2+} may be translocated to mitochondria and initiate pore opening.

Low concentrations of adenine nucleotides may be recognized by the ANT/VDAC complex and are thought induce pore opening [131]. It has been demonstrated that the binding of both ADP and ATP can inhibit pore opening induced by Ca^{2+} overload [130], suggesting that alteration of the adenine nucleotide concentration may be able to induce pore opening and apoptosis induction.

Intracellular *increase in P_i* occurs concomitantly with the net loss of adenine nucleotides and is known to enhance pore opening during Ca^{2+}-induced permeability transition [136].

Proteins of the Bcl-2 family are well known for their ability to regulate mPTP opening, probably by interacting with VDAC at the outer mitochondrial membrane. The anti-apoptotic proteins Bcl-2 and Bcl-X_L were shown to prevent mPTP-mediated membrane depolarization in intact cells as well as in isolated mitochondria [140, 141].

The pro-apoptotic protein Bax is thought to trigger pore opening by translocation to mitochondria, multimerization and binding to ANT [142, 143].

Bid, antother pro-apoptotic protein, is known to be responsible for the cross-talk between receptor-mediated and mitochondria-mediated apoptosis (chapter 2.1.). Bid may be cleaved to tBid (truncated Bid) by caspase-8, which then induces pore opening [116].

Interestingly, pore opening upon regulation by Bcl-2 proteins does not always include the ANT part of the pore. It was shown that the VDAC part of the pore opens, whereas the ANT part remains closed. This in turn does not result in the development of a pore permeable for molecules up to 1.5 kDa and the above described consequences. In contrast, a release of intermembrane proteins and subsequent apoptosis induction was observed without any loss of mitochondrial activity [144].

However, the detailed mechanisms of Bcl-2 proteins in the regulation of mPTP-opening are quite complex. [135, 143-146].

Low Δψ, as a result of uncoupling of the respiratory chain may lead to pore opening as demonstrated by the use of uncoupling agents in isolated mitochondria [147]. Due to its voltage dependent properties, VDAC is able to react to Δψ-loss and induce pore opening [148], but the associated mechanisms are rather speculative and there is no evidence that low Δψ alone can efficiently induce pore opening _in vivo_ [130].

Several other factors have been proposed to induce the opening of the mPTP, e.g. caspases, indicating a positive feedback loop in apoptosis [149] or changes in the pH value [130, 146] but the most potent inducers of pore opening seem to be ROS.

ROS – the most efficient inducers mPTP opening

Although pore opening can also be detected in systems lacking O_2 [146], ROS are probably the most dominant inducers of mPTP alterations. Either produced by mitochondria or by an external stimulus, increased intracellular ROS-levels have been shown to efficiently induce mPTP opening and apoptosis in various experimental setups [150-152]. The pore opening capacity of ROS seems to be enormous because it is able to (i) overcome the anti-apoptotic effect of Bcl-2 [153], (ii) override the pore-inhibitory action of intramitochondrial adenine nucleotides [130] and (iii) act independently of mitochondrial Ca^{2+} overload [154, 155].

Under normal circumstances, the decision whether apoptosis should be executed or not is weighed very properly and the mechanisms of pore opening are highly complex and fine-tuned. ROS, however, pose an evolutionary old and very harmful threat at the cell. If the intracellular ROS concentration reaches a level the cell can not cope with, the inductions of apoptosis has to be irrevocable in order to avoid ROS-mediated necrotic cell death which would lead to tissue inflammation [38, 105].

In the light of these facts, He et. al. proposed a new model of "unregulated" pore opening induced by severe oxidative stress, as apposed to the model of "regulated" pore opening induced by factors mentioned above [156]. During unregulated permeability transition mPTPs are thought to be opened due to permanent structural changes of the pore complex induced by ROS. The regulated mode of pore opening is mainly activated and controlled by Ca^{2+} and can be inhibited by cyclosporine A and Mg^{2+} whereas the unregulated mode is Ca^{2+}-independent and insensitive for cyclosporine A and Mg^{2+} [156].

A number of comprehensive reviews [105, 131, 146] have addressed ROS-mediated pore opening, yet the exact mechanisms are still not understood in full detail. Some suggestions and theories will be discussed in the following.

Involvement of lipid peroxidation in ROS-mediated pore opening

The primary target of ROS in mitochondria, apart from the pore complex itself, are lipids in the inner and outer membrane. It was shown that ROS-induced lipid peroxidation is followed by the opening of mPTPs [146]. Lipid peroxidation of mitochondrial membranes may evoke membrane destabilization processes which induces conformational changes inside VDAC-ANT part of the pore complex [146]. These changes subsequently lead to channel-like properties of the complex and the pore opens.

Another theory emphasizes the effect of mitochondrial lipid peroxidation on the mitochondrial transmembrane potential [157, 158]. Due to the oxidative action of ROS, the lipid layer of the inner mitochondrial membrane gets leaky and H^+ can pass. As a consequence, $\Delta\psi$ drops significantly which is thought to increase the probability of pore opening.

Involvement of Ca^{2+} overload in pore opening

Oxidative stress is often accompanied by a critical increase in intracellular Ca^{2+}. There is evidence that ROS are always required for Ca^{2+}-mediated pore opening and apoptosis [131, 155, 156, 159]. ROS mediated damage to the endoplasmatic reticulum has been shown to result in Ca^{2+} influx into mitochondria which in the following induces permeability transition and apoptosis induction [160, 161]. Kowatolwski et. al. suggested that Ca^{2+} ions amplify the membrane depolarization induced by ROS by binding and destabilization of lipids on the inner surface of the mitochondrial membrane [139, 162]. Other studies have shown that Ca^{2+}- and P_i-induced pore opening can be partially blocked by catalase, indicating the involvement of H_2O_2 [163]. In contrast, some researchers propose that Ca^{2+} overload is only a consequence of a bioenergetic failure that occurs after the mPTP-opening and is not a indispensable factor for the onset of permeability transition [155].

ROS-mediated damage to Bcl-2 and pore opening

Cytosolic ROS can damage Bcl-2 and may suppress its anti-apoptotic activity and therefore enhance the probability of mPTP opening and apoptosis [164]. PDT applications producing cytosolic ROS have been shown to induce apoptosis by permanent mPTP opening, even if Bcl-2 is overexpressed [165].

Oxidation of critical SH residues in the pore complex induce pore opening

The most likely mechanism for ROS to induce pore opening is the oxidation of critical parts in the structure of the protein complex, leading to conformational changes and pore opening.

As discussed in chapter 1.3., ROS damage proteins by oxidizing SH-residues which leads to the generation of disulfide (S-S) bridges inside the proteins structure. This is usually followed by conformational changes in the proteins tertiary structure, thus representing a mechanism that perfectly fits for ROS-mediated pore opening.

Indeed, diamide and other thiol crosslinkers have been shown to induce pore opening in isolated mitochondria as well is *in vivo* suggesting pore opening due to S-S mediated conformational changes inside the protein complex [166-168].

The translocator protein ANT contains an SH-group of a cystein residue (Cys56) which is structurally exposed to ROS-mediated damage may therefore be easily converted to form a disulfide bridge. Exactly this conversation has been shown to be induceable by thiol crosslinkers and ROS in intact cells, isolated mitochondria and in purified ANT molecules and is assumed to be responsible for irreversible pore opening and apoptosis induction [138, 146, 153]. This assumption is emphasized by the finding that the addition of disulfide reductants like dithiotreitol (DTT) could inhibit pore opening by ANT modification [169].

Another study provides evidence that the oxidation of Cys56 in ANT induces apoptosis beyond any control of Bcl-2 or Ca^{2+} [153], supporting the theory of He et. al. of ROS-mediated "uncontrolled" pore opening.

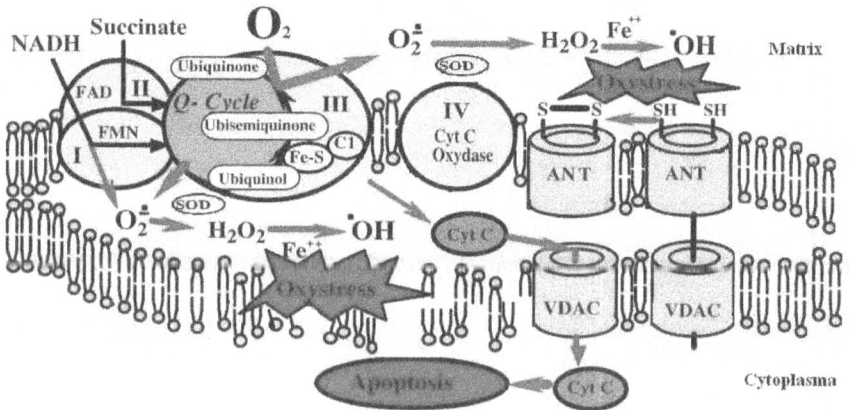

Figure 2.5. ROS-mediated opening of the mPTP. Mitochondrial derived ROS oxidize (i) crictical SH residues in ANT and (ii) the lipid layer of mitochondrial membranes, which both result in the opening of the mPTP.

ROS-induced pore opening and subsequent apoptosis as a line of antioxidant defense

Depending on the extent of ROS overproduction, ROS-mediated opening of mPTPs can have different consequences for the cell. According to Skulachev, several cellular strategies act against the threat of ROS [4, 104, 105].

(i) ROS overproduction initially induces lipid peroxidation of the inner mitochondrial membrane, leading to *mild-uncoupling* - the *first line of antioxidant defense* (see chapter 1.4.).

(ii) If mild-uncoupling is insufficient to stop ROS overproduction, *short-term opening of mPTPs* ensues as the *second line of antioxidant defense*. Due to the opening of the mPTPs, respiratory phosphorylation is now completely uncoupled. As immediate response, the rate of respiration is increased to its maximum, leading to excessive O_2 consumption. Since all present O_2 molecules are used for respiratory processes, the overall intracellular O_2 level is actively decreased. In addition, the lifetime of $^\bullet Q^-$ is minimized, decreasing that the probability of any unwanted one-electron transfer. Furthermore, small amounts of cyt.c may be released to cytosol during short term pore opening in order to scan for $^\bullet O_2^-$. As described in chapter 1.4., cyt.c is able to directly detoxify $^\bullet O_2^-$ and may therefore act as an antioxidant defense mechanism.

All these mechanisms lead to ROS decrease, which was observed to be followed by the closing of mPTPs [130, 146].

With regard to the fact that ROS are thought to induce pore opening by permanent modifications of ANT, it is justified to question how reversible pore opening is possible even if the intracellular ROS-concentration decreases.

One possible answer could be the action of natural disulfide reductants such as thioredoxins (Trx). Trx binds to target proteins and reduces their disulfide bridges. In this manner, thioredoxins might act to repair modifications of Cys56 in the ANT while ROS generation decreases. Accordingly Trx-overexpression has been shown to block mitochondrial apoptosis in leukemia cells [170].

Another theory is that a moderate ROS overproduction may not be sufficient to oxidize the ANT, but to be potent enough to induce lipid peroxidation. As discussed above, lipid peroxidation may have different effects which finally lead to pore opening. As a countermeasure, the enzymatic activity of phospholipase A may be sufficient to repair the damage done to the mitochondrial membrane as soon as the ROS concentration decreases again, thus favoring pore closing [130, 138].

However, the detailed mechanisms of ROS-mediated temporary mitochondrial permeability is still not fully understood.

(iii) If ROS still accumulate inside the mitochondria, mitochondrial permeability becomes irreversible and a permanent osmotic disbalance between matrix and intermembrane space occurs, eventually leading to the destruction of the mitochondria and release of intermembrane proteins into the cytosol – a process termed *mitoptosis* [38, 106]. Mitoptosis is regarded as a form of isolated "organelle-suicide" that is not necessarily followed by apoptosis of the whole cell. Mitoptotic cells may survive after they have eliminated malfunctioning mitochondria that are overreactive to ROS, thus representing the *third line of antioxidant defense*.

(iv) If the majority of mitochondria inside the cell are subjected to mitoptosis, all preapoptotic antioxidant measures have failed. As an irreversible consequence, a high concentration of intermembrane proteins gains access to the cytosol and the *apoptotic machinery* is initiated, leading to the elimination of ROS-over producing cells. With regard to Skulachev's theory, this represents the *fourth line of oxidative defense*, because it serves as a means by which the organism can get rid of potentially harmful cells before they in turn can affect their neighbouring cells.

The orchestrated actions of cellular antioxidant defense, mitoptosis and, lastly, apoptosis have been summarized in a comprehensive review the title of which emphasizes the major principle behind all the defensive measures of the cell against ROS overproduction – "*It is better to die than to be wrong*", for which Skulachev also coined the term "Samurai law of biology" [38].

In this context, even the term "phenoptosis" has been proposed as a description of the programmed death of a whole organism for the "collective good", exemplified for instance by the death of the salmon after it has spawned [105, 171].

2.3. Antioxidants, apoptosis and cancer

It is a well known matter of fact that antioxidants can protect the organism against serious mutations that can cause the development of malign tumors [172, 173]. Especially exogenous antioxidants which enter the organism via nutrition, e.g. ascorbic acid, α-tocopherol and carotenoids, are regarded to prevent cancer development and are often given as a co-treatment during chemotherapy [174].

Diets supplemented with β-carotene, vitamin E and selenium have been observed to decrease the rate of esophageal cancer [175]. α-tocopherol and β-carotene have been described to delay the development of gastric cancer [176] and decreased the risk of colon [177] and prostate cancer [178]. These antioxidant applications are supported by *in vitro* studies and animal models that demonstrate a clear relationship between oxidative damage to DNA and the onset of tumor genesis [179].

Despite their beneficial action, antioxidants sometimes fail to prevent tumor development. In some cases antioxidant-supplemented nutrition may even enhance tumor growth instead of suppressing it [180]. For example, diet supplementation with α-tocopherol and β-carotene was not able to prevent the occurrence of lung cancer among male smokers. Even more surprisingly, persons who were supplemented with β-carotene exhibited a higher incidence of lung cancer and mortality than control persons [180, 181].

Another study, which used a well known tumor model in mice, demonstrated that antioxidant depletion did not, as expected, enhance tumor growth, but significantly inhibited tumor progression [182].

In the light of these observations, one may ask how alteration of the antioxidant capacity can have such different effects on tumor development and tumor growth. The answer to this question might be given by using the knowledge about ROS and apoptosis induction.

The assumption that antioxidants can protect from cancer development mediated by oxidative stress may be true for healthy individuals without any profound cancer-risks but not for people that already suffer from cancer or that are continuously exposed to high concentration of carcinogens.

Heavy smokers, for example, constantly face the threats of chemical carcinogens and ROS evoked by the daily use of tobacco. Therefore lung tissue inevitably harbors numerous mutagenized precancerous cells that have to be eliminated by apoptosis in order to avoid the development of malign cancer cells. A use of antioxidants may reduce the damage set by smoke-derived ROS but not the one set by chemical carcinogens. Normally, the oncogen p53 recognizes DNA-damage in precancerous cells and induces apoptosis by mechanisms which involve ROS overproduction. If the intracellular level of antioxidants is too high, the ROS level that is required to efficiently execute apoptosis cannot be maintained. This could lead to a survival of potentially malign cells and therefore to an increased probability of cancer development [180]. In contrast, non-smokers in whom lung-tumor genesis is not dependent on chemical carcinogens but on metabolic derived ROS can decrease the probability of cancer-development by frequently using antioxidant supplements [183].

Using a defined brain tumor model with known p53-mediated apoptosis rates in transgenic mice Salganik et. al. compared the impact of antioxidant-depleted to an antioxidant-enriched diet on cancer progression [182]. Dramatically inhibited cellular proliferation due to enhanced apoptosis was observed in tumor tissue, but not in normal tissue of antioxidant-depleted mice. Together with the fact that increased oxidative stress in tumor tissue could be detected, this lead the authors to conclude that ROS are responsible for the observed tumor reduction.

With regard to the above-summarized findings the following key points concerning antioxidant supply in humans during cancer therapy and as cancer-prevention tool may be stated [180]:

(i)   Due to genetic and environmental factors the human population is heterogeneous in endogenous levels of ROS, resulting in a heterogeneous relevance for the use of antioxidants.

(ii)   People who overproduce ROS due to some pathological disturbances are at high risk for developing cancer and other serious diseases. These people should be encouraged to increase their antioxidant capacity by the daily use of antioxidant supplements.

(iii)   People with a low level of ROS or people exposed to other sources of carcinogenesis should avoid the excessive use of antioxidants. A certain level of intracellular ROS generation has to be maintained in order to enable highly important protective reactions like apoptosis, deleting precancerous, cancerous, virus infected or other cells which threaten human health.

(iv)   Screening the human population for ROS levels can provide a scientifically well-based, controlled application of antioxidants and might significantly contribute to improve human health.

The observed alteration of "natural" apoptosis in cancerous cells evoked by modulated antioxidant capacity leads to another question:

Do antioxidants inhibit the cytotoxic effect of anti-cancer applications that depend on ROS-induced cell death? Or: Can the cytotoxic effect be enhanced by lowering the cells antioxidant capacity?

Cisplatin, for example, is used as an anticancer drug because of its ability to efficiently induce apoptosis by generation of ROS. Pre-incubation of breast cancer cells with α-tocopherol has been shown to inhibit cisplatin-mediated ROS generation and subsequent apoptosis [180].

Furthermore, the addition of BSO to bladder cancer cells was shown to deplete intracellular GSH, leading to enhanced ROS generation and cell death induced by cisplatin [184].

The PKC inhibitor staurosporine (STS), a bacterial alkaloid compound, is known to induce apoptosis in various cells types, including cancer cells. Due to these properties, STS and its derivates are widely used in chemotherapeutic applications [185, 186]. It has been proposed in several studies that ROS are the mediators of STS-induced mitochondrial permeability transition and apoptosis [186-189]. For instance, Santamaria et. al. have demonstrated that STS induced apoptosis in rat hepatoma cells is mediated by ROS overgeneration at complex I of the respiratory chain, most likely via reverse electron flow [190].

The effects of antioxidants on STS-mediated ROS production may be substantial.

Cytosolic and mitochondrial ROS have been identified to be responsible for STS-induced apoptosis in retinal cells and the addition of several artificial antioxidants (like GSH-ee, Trolox of MnTBAP) can significantly reduce ROS generation and apoptosis, indicating a primary role of ROS in the induction of apoptosis induced by STS [150].

STS-induced ROS production and apoptosis in cortical neurons has been shown to be attenuated by the addition of synthetic SOD/Cat mimetics of the Euk-family suggesting that superoxide might be the reactive species which is mainly produced by STS [191].

Photodynamic therapy induces cell death via production of mitochondrial and cytosolic ROS, respectively. It has been demonstrated in numerous studies that antioxidants can influence the outcome of PDT.

In an early study Miller et. al. have shown that intracellular depleted GSH (either by the addition of BSO or by genetic manipulations) shifted the cellular toxicity to lower PDT-irradiation, an effect observed in four different cell lines [93].

A combination of PDT and BSO-treatment in rats with intracerebral tumors enhances the cytotoxic effect of PDT and cell death appears to be shifted from an apoptotic phenotype to a necrotic one [88]. The same effect has been reported for PDT/BSO combination treatment of gliosarcoma tumors in rat brains [88].

A protective effect of GSH in ALA-PDT has been reported by Kliukiene and coworkers [89]. The addition of erythrocyte-glutathione reductase in kidney fibroblast cell culture has been demonstrated to diminish the cytotoxic effects of PDT in healthy, tumor-surrounding cells and moreover, minimizes the negative side effects, like immoderate necrosis.

This protective effect of GSH and other free radical scavengers in PDT could be verified by Perotti and coworkers [192].

III. Aims of the studies

Although all three individual studies presented in this doctoral thesis address different specific topics, they are linked together by one question:

How may the intracellular antioxidant capacity influence the outcome of ROS-mediated cancer therapies, i.e. PDT and Ro-31-8220 treatment?

I: Photodynamic Treatment with Fractionated Light Decreases Production of Reactive Oxygen Species and Cytotoxicity *in vitro* via Regeneration of Glutathione

Photodynamic therapy (PDT) is a specific type of tumor therapy that is used to remove unwanted or harmful cells by intracellular overproduction of reactive oxygen species (ROS). PDT treatment consists of application of a photoactive compound to a selected tissue and irradiation of the compound with light of the appropriate wavelength. In tumor tissue with insufficient blood supply, where the oxygen status is very low, light application during PDT is sometimes delivered in a fractionated manner to permit the reoxygenation of the treated area in order to maximize the tumoricidal effect.

This study was aimed at investigating the outcome of fractionated irradiation in an *in vitro* PDT system, using Hypericin and AlPcS$_4$ as photosensitizers and the human epidermoid carcinoma cell line A431 as cell model. In this model system deoxygenation can be neglected because oxygen is present throughout the entire culture and treatment period. Fractionated irradiation with light/dark intervals of 45/60 seconds was found to decrease ROS production and thus the cytotoxicity of PDT in comparison to continuous irradiation. We thus wondered whether antioxidant mechanisms, particularly the GSH antioxidant system, may be involved in the effects observed with fractionated irradiation. Indeed, the attenuated cytotoxic outcome of fractionated PDT *in vitro* could be reversed by addition of BCNU, an inhibitor of the glutathione reductase, suggesting that the dark intervals during fractionated irradiation allow for the glutathione reductase to regenerate GSH, thereby rendering cells less susceptible to ROS produced by fractionated irradiation compared to continuous irradiation.

These findings provide additional information about the mechanisms of PDT and may be of particular interest for clinical PDT applications where irradiation is applied to differentially oxygenated tumors.

II: Differential Effects of Glucose Deprivation on the Cellular Sensitivity Towards Photodynamic Treatment-Based Production of Reactive Oxygen Species and Apoptosis Induction

Apoptosis is an energy-consuming process that is critically dependent on continuous supply with ATP produced by the mitochondrial respiratory chain. Loss of the mitochondrial membrane integrity has been shown to result in a serious disturbance of the intracellular ATP supply. To a certain extent, this effect may be compensated by glycolysis, which is able to produce ATP in a sufficient quantity in order to maintain the capacity for apoptosis induction and execution. One of the most potent inducers of mitochondrial membrane damage and thus apoptosis are reactive oxygen species (ROS) which have been described to be produced in response to photodynamic therapy (PDT). Correspondingly, the apoptosis-inducing capacity of PDT is known to be critically influenced by cellular antioxidants.

Based on this knowledge, the present study addresses the effects of altered glycolysis in human epidermoid carcinoma cells (A431) on the cellular GSH antioxidant system and the subsequent apoptosis induction by PDT, using AlPcS$_4$ as photosensitizer.

We found that disruption of glycolysis by glucose deprivation prior to PDT caused a reduction of the intracellular GSH concentration, thus decreasing the function of an important member of the antioxidant defense against PDT-mediated ROS production. Similarly, addition of artificial inhibitors of GSH synthesis and regeneration (BSO and BCNU, respectively) resulted in the enhancement of PDT-mediated apoptosis. Interestingly, impairment of GSH biosynthesis by glucose deprivation induced a shift of the cell death mode from apoptosis to necrosis, indicating that glucose deprivation not only prevented GSH biosynthesis but also caused a profound deficiency in ATP production.

These observations suggest that in situations where mitochondrial damage is accompanied by impaired glucose supply, PDT preferentially results in necrotic cell death because the energy produced by glycolysis that is required for efficient apoptosis induction and execution is no longer available.

These results contribute to the understanding of how the cellular energy metabolsim may influence the outcome of PDT and provide important clues about the involvement of antioxidants in various forms of therapeutic applications that are based on apoptotic cell clearance.

III: The staurosporine analoge Ro-31-8220 induces mitochondrial apoptosis by overproduction of reactive oxygen species

Agents with the potential to inhibit protein kinase C (PKC) have been successfully employed as effective tools in anticancer therapy because of their ability to modify cell cycle, proliferation and apoptosis. Ro-31-8220, a derivative of staurosporine (STS), is one such agent and shares with STS the ability to efficiently induce apoptosis. While STS-mediated apoptosis has been evidently linked to increased intracellular production of ROS, the mechanisms of Ro-31-8220-induced apoptosis have not yet been definitely determined.

This study was aimed at investigating the involvement of ROS in Ro-31-8220-induced apoptosis in A431 human epidermoid carcinoma cells. Ro-31-8220 application was found to drastically increase intracellular ROS levels during the early phase of treatment and trigger a cascade of apoptotic events that ultimately lead to cell death. Interestingly, Ro-31-8220-mediated ROS production appeared to be dependent on the integrity of mitochondrial membranes, indicating that the largest part of ROS is generated inside mitochondria.

In order to investigate whether the observed ROS production is initially responsible for effective apoptosis induction by Ro-31-8220, the influence of several artificial antioxidants was tested. As a result, ROS production and apoptosis could be significantly inhibited or even completely suppressed by pre-treatment with antioxidants, indicating that ROS are indeed responsible for apoptosis initiation by Ro-31-8220.

Taken together, the results obtained in this study clearly demonstrate that Ro-31-8220 induces apoptosis in A431 cell *via* production of intracellular ROS. This cytotoxic effect can be attenuated by enhancing intracellular antioxidant capacity, a fact that has to be considered when using Ro-31-8220 as chemotherapeutic agent combined with antioxidant co-treatment.

IV. Photodynamic Treatment with fractionated Light Decrease Production of Reactive Oxygen Species and Cytotoxicity *in vitro via* Regeneration of Glutathione

Photochemistry and Photobiology, 2005, 81: 609–613

Research Note

Photodynamic Treatment with Fractionated Light Decreases Production of Reactive Oxygen Species and Cytotoxicity *In Vitro via* Regeneration of Glutathione[¶]

Christian Benno Oberdanner, Kristjan Plaetzer, Tobias Kiesslich and Barbara Krammer*

Division of Allergology and Immunology, Department of Molecular Biology, University of Salzburg, Hellbrunnerstrasse 34, 5020 Salzburg, Austria

Received 23 August 2004; accepted 19 January 2005

ABSTRACT

Photodynamic therapy removes unwanted or harmful cells by overproduction of reactive oxygen species (ROS). Fractionated light delivery in photodynamic therapy may enhance the photodynamic effect in tumor areas with insufficient blood supply by enabling the reoxygenation of the treated area. This study addresses the outcome of fractionated irradiation in an *in vitro* photodynamic treatment (PDT) system, where deoxygenation can be neglected. Our results show that fractionated irradiation with light/dark intervals of 45/60 s decreases ROS production and cytotoxicity of PDT. This effect can be reversed by addition of 1,3-bis-(2-chloroethyl)-1-nitrosurea (BCNU), an inhibitor of the glutathione reductase. We suggest that the dark intervals during irradiation allow the glutathione reductase to regenerate reduced glutathione (GSH), thereby rendering cells less susceptible to ROS produced by PDT compared with continuous irradiation. Our results could be of particular clinical importance for photodynamic therapy applied to well-oxygenated tumors.

INTRODUCTION

Photodynamic treatment (PDT) is based on the administration of a photosensitizing agent followed by irradiation with light of an appropriate wavelength. This activates the photosensitizer, which is further able to react with molecular oxygen ($^1O_2$), generating either singlet oxygen ($^1O_2$) or other reactive oxygen species (1,2), eventually leading to necrotic and apoptotic cell death (3,4).

As demonstrated in several studies, photodynamic therapy effects may be augmented *in vivo* by modifying the irradiation regime without increasing the total light energy (J/cm^2) by either modulating the fluence rate (5,6) or by introducing dark intervals during light application (fractionated irradiation) (6–8).

Under conventional clinical photodynamic-therapy conditions, the amount of molecular oxygen in the treated area is rapidly depleted during the photochemical processes, leading to a diminished capacity for further production of reactive oxygen species (ROS) (8–10). This effect is especially important in antitumor applications of photodynamic therapy where the tumor areas are characterized by poor vascularization or where photodynamic therapy causes damage to afferent blood vessels, both leading to insufficient oxygen supply during treatment (11,12). Deoxygenation may be reduced by fractionated light delivery by providing sufficient time for reoxygenation of the treated tissue during the dark periods (6,8). Additional antitumor effects of fractionated irradiation are ROS damage produced during reperfusion of the area, which occurs after a transient period of ischemia (reperfusion injury) (13) and relocalization of the sensitizer during the dark interval(s) (14).

Apparently, the success of fractionated photodynamic therapy is dependent on several factors, with the light-delivery regime as the one probably most important (9,10,15). Several theoretical and practical approaches were able to demonstrate that the most effective light fraction schedules consist of equal light on and off periods, with intervals between 15 and 60 s (9,10). Shorter intervals have been shown to cause other mechanisms to become dominant, as for intervals in the nanosecond levels an enhanced photodynamic effect may occur (16).

However, the literature on fractionated light delivery in photodynamic-therapy applications is still contradictory and incomplete, especially for *in vitro* systems. Some investigators did not find any positive antitumor action of fractionated photodynamic therapy *in vivo*, which may be explained by the influence of cellular repair mechanisms becoming active in the dark intervals to compensate for the initial damage (6,17).

In the present study, we addressed the question whether regeneration of ROS quenching mechanisms, namely the glu-

¶Posted on website on 1 February 2005

*To whom correspondence should be addressed: Division of Allergology and Immunology, Department of Molecular Biology, University of Salzburg, Hellbrunnerstrasse 34, 5020 Salzburg, Austria. Fax: ++43-662-8044-150; e-mail: barbara.krammer@sbg.ac.at

Abbreviations. AlPcS4, aluminum (III) phthalocyanine tetrasulfonate; BCNU, 1,3-bis-(2-chlorethyl)-1-nitrosurea; DCF, 2'-7'-dichlorofluorescein; DCFH-DA, 2',7'-dichlorodihydrofluorescein diacetate; DMEM, Dulbecco's modified Eagle's medium; FCS, fetal calf serum; GSH, reduced form of glutathione; GSSG, oxidized form of glutathione; MTT, 3-(4,5-dimethyl-2-thiazolyl)-2,5-diphenyl-2H-tetrazoliumbromide; NADPH, reduced form of nicotinamide adenine dinucleotide phosphate; PBS, phosphate-buffered saline; PDT, photodynamic treatment; ROS, reactive oxygen species.

610   Christian B. Oberdanner *et al.*

tathione system, influences the outcome of fractionated PDT in an *in vitro* system, where oxygen supply during PDT is granted. We, therefore, monitored ROS formation and the cytotoxic effect of fractionated PDT, using hypericin and aluminum (III) phthalocyanine tetrasulfonate (AlPcS$_4$) as sensitizers, with and without pharmacological inhibition of glutathione regeneration.

MATERIAL AND METHODS

Cell culture and photodynamic treatment. A431 human epidermoid carcinoma cells (ATCC CRL-1555) were cultured in Dulbecco's modified Eagle's Medium (DMEM) containing 4.5 g/L glucose supplemented with 10 mM 2-[4-(2-hydroxyethyl)-piperazin-1-yl]ethanesulfonic acid (HEPES), 4 mM L-glutamine, 1 mM Na-pyruvate, 100 U/mL penicillin, 0.1 mg/mL streptomycin and 5% fetal calf serum (FCS) (all from PAA-laboratories, Linz, Austria), in a humidified atmosphere at 37°C and 7.5% CO$_2$. For measurement of ROS production, 3×10^5 cells in 1.5 mL 5% FCS DMEM were seeded into 30 mm Petri dishes (Greiner Bio-One, Kremsmuenster, Austria); for measurement of cytotoxicity, 7500 cells in 100 µL 5% FCS DMEM were seeded into 96-well microplates (black walls, clear bottom; Greiner Bio-One). In a first step, cells were incubated for 20 h and 18 h in 0% FCS DMEM containing 1.5 µg/mL hypericin (Fluka BioChemika, Buchs, Switzerland) and 10 µM AlPcS$_4$ (Porphyrin Products, Logan, UT), respectively. For inhibition of the glutathione reductase, supernatants were aspirated and fresh, sensitizer-free medium containing 500 µM 1,3-bis(2-chloroethyl)-1-nitrosourea (BCNU; Sigma-Aldrich, Vienna, Austria) (18) was added for an additional hour.

Immediately before irradiation, the culture supernatants of all samples were replaced with fresh media. Irradiation was performed using a red-light irradiation diode array with λ_{max}, AlPcS$_4$ = 660 ± 20 nm (cat. no. L53SRC-F) and a light power of 10 mW/cm^2 and λ_{max}, hypericin = 610 ± 20 nm (cat. no. L7113SEC), 0.9 mW/cm^2, respectively (both superbright diodes, Kingbright Electronics, Issum, Germany). Fractionated light was delivered in a discontinuous irradiation regime, consisting of 45 s irradiation and 60 s dark intervals.

Samples were protected from ambient light to avoid unspecific sensitizer activation; for all experiments, cells from passages 5–15 were used.

Measurement of cytotoxicity. Metabolic/mitochondrial activity was assessed as before (19,20) 24 h post irradiation by the reduction of 3-(4,5-dimethyl-2-thiazolyl)-2,5-diphenyl-2 H-tetrazolium bromide (MTT; Sigma-Aldrich) to the insoluble blue formazan catalyzed by mitochondrial dehydrogenases (21).

Measurement of intracellular ROS. Intracellular ROS were measured by the oxidation of 2',7'-dichlorodihydrofluorescein diacetate (DCFH-DA; Molecular Probes Europe, Leiden, Netherlands) and quantified by flow cytometry (FACS Calibur, Becton-Dickinson, Franklin Lakes, NJ) (22,23).

Thirty minutes prior to irradiation, cells were incubated with 50 µM DCFH-DA and PDT-treated as described above. Ten minutes after irradiation, cells were harvested together with the supernatants using Accutase (PAA-laboratories); all subsequent steps were performed on ice. The cells were spun down at 840 g at 4°C, washed with 1 mL phosphate-buffered saline (PBS), the pellets were resuspended in 500 µl PBS, the resulting green fluorescent DCF (2'-7'-dichlorofluorescein) was analyzed by flow cytometry (Fl-1 channel; λ_{EX} = 488 nm, λ_{EM} = 530 ± 30 nm). For each sample, 10000 cells were analyzed and the fluorescence signals were evaluated as described in the *Results* section.

Statistical evaluation. Data represent mean of three independent experiments ± SEM. Statistical significance was evaluated using Student's *t*-test.

RESULTS

Cytotoxicity after PDT with continuous and fractionated irradiation

In order to determine the viability of A431 cells following PDT with AlPcS$_4$ and hypericin under either continuous or fractionated irradiation conditions, we used the MTT assay to measure the metabolic/mitochondrial activity. To specifically investigate the impact of the reduced form of glutathione (GSH) recycling system

on the viability of the treated cells, we added BCNU, an inhibitor of the glutathione reductase, interfering with the cells' capacity to reduce the oxidized form of glutathione (GSSG) to its reduced form, GSH. In control experiments, we could show that BCNU application for 1 h *per se* was not toxic within the first 24 h after addition (data not shown) and, furthermore, did not affect the intracellular GSH levels (see [24]). Additionally, no cytotoxic effect was found for dark controls (only photosensitizer without irradiation) and light-only controls (irradiated without sensitizer) (data not shown).

Figure 1 shows the percentage of MTT signal under several conditions after AlPcS$_4$-PDT (A) and hypericin-PDT (B), related to the respective controls (not illuminated but identically treated samples without sensitizer).

The MTT signal for AlPcS$_4$-sensitized and continuously irradiated cells (filled squares) was reduced quite rapidly to <5% in a fluence range from 1.5 to 3.5 J/cm^2. The curve of fractionatedly irradiated samples (filled triangles) is similarly shaped, but the cells were far less sensitive to photodynamic treatment. The MTT signal remained at the control level up to a light dose of approximately 2 J/cm^2 and eventually decreased to about 15% at a light dose of 4 J/cm^2.

Expectedly, BCNU-treated cells appeared to be more sensitive to PDT-mediated cell killing than those without BCNU. Cells treated with continuous PDT and BCNU (open squares) showed a rapid decline of the MTT signal, starting at a light dose of about 1 J/cm^2 and reaching <5% at approximately 3 J/cm^2. Addition of BCNU (open triangles) seemed to compensate for the effect of fractionated irradiation at least at fluences >2.25 J/cm^2; the survival curve of cells treated with fractionated irradiation and BCNU and the curve of continuously irradiated cells appear similar, both reaching the base level at about 3.5 J/cm^2.

The results of hypericin experiments in principle match those of AlPcS$_4$-PDT; the particular curves are similarly shaped, albeit different fluences were applied.

ROS production after PDT with continuous and fractionated irradiation

In order to explore whether the decreased cytotoxicity of fractionated light application was caused by lower ROS formation, we determined intracellular ROS levels in A431 cells after PDT under the conditions mentioned above. In control experiments, no increase of the DCF signal was found in dark controls (only photosensitizer without irradiation) as well as in light-only controls (irradiated without sensitizer) (data not shown).

Figure 2 shows the percentage of the DCF signal after AlPcS$_4$-PDT (A) and hypericin-PDT (B) with varying light doses, related to respective control cells (not illuminated, without AlPrS$_1$ and either with or without BCNU); addition of BCNU increased the ROS level of controls by a factor of about 2.5 due to a reduced capacity of GSH to detoxify endogenously generated ROS (24).

In AlPcS$_4$-PDT, the dynamics of ROS levels generated by continuously irradiated cells were characterized by an increase of the ROS content from about three-fold of control levels at a light fluence of 0.9 J/cm^2 to the 6.5-fold at a light fluence of about 1 J/cm^2. In contrast, light delivery in fractions caused a lower oxidation of DCFH-DA, starting from approximately two-fold (0.9 J/cm^2) to four-fold (1.8 J/cm^2) of control.

The addition of BCNU provoked corresponding effects to those seen in the cytotoxicity experiments (Fig. 1). Continuous irra-

Photochemistry and Photobiology, 2005, 81   611

diation together with addition of BCNU caused a very rapid increase of intracellular ROS from approximately 5.5-fold at a light dose of 0.9 J/cm^2 to eight-fold at 1.8 J/cm^2. Treatment with BCNU reversed the effect of fractionated light application; the relative ROS levels increased to about five-fold and 7.5-fold with light doses of 0.9 and 1.8 J/cm^2, respectively.

In hypericin-PDT, the results fully confirm those of AlPcS$_4$-PDT at slightly lower overall levels of ROS compared with AlPcS$_4$-PDT experiments.

DISCUSSION

In vivo fractionated light delivery in photodynamic therapy causes, in many cases, increased cytotoxicity, which can mainly be explained by reoxygenation effects during the dark periods (6,8,9,25). Additionally, relocalization of the particular photosensitizing agent (14) and induction of apoptosis by reperfusion injury (13) have been reported to occur and support the success of photodynamic therapy. Effective light fraction schedules have been shown to consist of equal light and dark periods with intervals between 15 s and 1 min (9,10).

However, this time frame is sufficient to restore the ROS quenching capacity of the glutathione system by the fast-acting, reduced form of nicotinamide adenine dinucleotide phosphate (NADPH)-dependent enzyme glutathione reductase (18,26,27). The latter antagonizes the PDT-induced damage to a certain degree, depending on several parameters, such as the dominant reactive oxygen species or the localization of the photosensitizer, both of which might have a negative impact on the glutathione reductase itself.

Within this study, we could show that, under *in vitro* PDT conditions, where sufficient oxygen supply due to the high amount of oxygen in the supernatant medium and continuous gas exchange with the outside is granted, fractionated irradiation using light/dark intervals of 45/60 s was found to reduce intracellular ROS levels, compared with continuous irradiation, by a factor ranging from 1.5 and 1.1 at low fluence rates to 1.7 and 1.2 at high fluences, for AlPcS$_4$- and hypericin-PDT, respectively. Accordingly, the cytotoxic effect decreased when irradiation was performed in fractions. Addition of BCNU, an inhibitor of the glutathione reductase (18), reversed the effect of fractionated irradiation to a large extent. Both effects, lower ROS levels and decreased cytotoxicity, and their reversibility by BCNU could be observed for two photosensitizing agents with different chemical properties, *i.e.* the hydrophilic AlPcS$_4$ (mainly localizing in lysosomes [28,29]), and the lipophilic hypericin (localizing in mitochondria and membrane systems [30,31]). These different chemical properties ultimately reflect the ROS production induced by PDT with the respective sensitizer: AlPcS$_4$-mediated PDT is known to preferentially produce hydroxyl radicals (32) whereas hypericin-PDT more likely induces the formation of singlet oxygen (33,34). This is of particular importance for the interpretation of the ROS measurements because DCFH-DA is specific for hydroxyl radicals and, only to limited degree, for other ROS (35,36).

Because GSH is mainly effective in quenching hydroxyl radicals (26), the effects of fractionated irradiation and BCNU addition on the production of ROS and cytotoxicity are more profound with AlPcS$_4$-PDT compared with hypericin-PDT.

Under conditions of continuous irradiation, samples incubated with BCNU produced more ROS than those without BCNU. This might be due to some regeneration of GSH (and subsequent de-

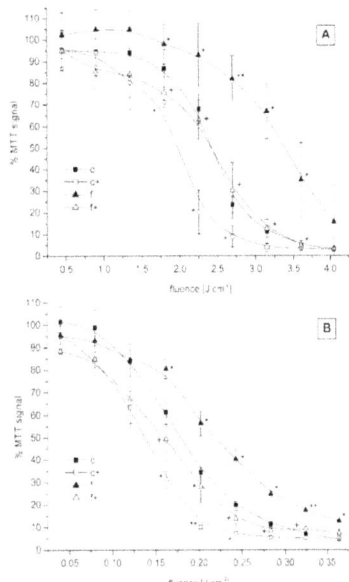

Figure 1. Metabolic/mitochondrial activity of AlPcS$_4$- (A) and hypericin-(B) photosensitized A431 cells after fractionated or continuous irradiation and/or under conditions of impaired glutathione recycling. Results are expressed as relative MTT activity related to control samples (without sensitizer and nonirradiated). Each point represents the mean of at least three independent experiments ± SEM. Data were statistically compared with continuously irradiated cells (**: $P < 0.01$, *: $P < 0.05$) and to fractionatedly irradiated samples (++: $P < 0.01$, +: $P < 0.05$), respectively. A: AlPcS$_4$, B: hypericin; (c) continuous, (c+) continuous + BCNU, (f) fractionated, (f+) fractionated + BCNU.

composition of ROS) in samples without BCNU under continuous irradiation, at least unless glutathione reductase is damaged by ROS, a possible occurrence at higher fluences. This effect is also reflected in the cytotoxicity measurements. The difference between continuous and fractionated irradiation is the (more or less) complete regeneration of GSH during the dark intervals in the latter.

We noticed a discrepancy in the ROS level and cytotoxicity for the fractionated and continuous irradiation with BCNU: the (statistically nonsignificant) differences in the DCF-signal result in a significant increase in cytotoxicity (at least for some doses, for AlPcS$_4$: results of statistical analysis not shown in Fig. 1). We suggest two hypotheses to explain this effect: (i) singlet oxygen (or other ROS not detected by DCFH-DA) is decomposed during the dark intervals of fractionated irradiation, thus leading to less cytotoxicity at similar DCF-signal levels; and (ii) cellular repair mechanisms restore the cellular protein/lipid ensemble during the dark intervals and thus cause less cytotoxicity.

612   Christian B. Oberdanner *et al.*

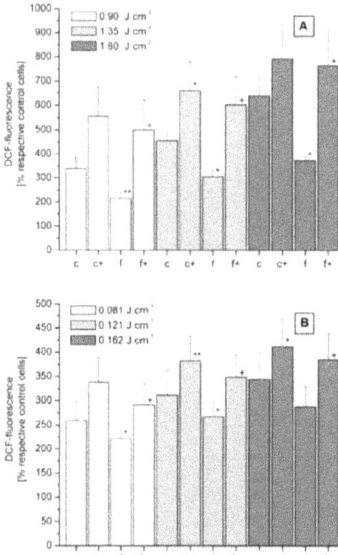

Acknowledgements—This study was supported by a research grant of the deanery of the natural sciences faculty, University of Salzburg, Austria (ad 1742/2003). The authors would like to thank Mag. Katrin Flatscher for proofreading the manuscript.

REFERENCES

1. Foote, C. S. (1990) Chemical mechanisms of photodynamic action. *Proc. SPIE Institute Advanced Optical Technologies on Photodynamic Ther.* **6**, 115–126.
2. Henderson, B. W. and T. J. Dougherty (1992) How does photodynamic therapy work? *Photochem. Photobiol.* **55**(1), 145–157.
3. Plaetzer, K., T. Kiesslich, T. Verwanger and B. Krammer (2003) The modes of cell death induced by PDT: an overview. *Med. Laser Appl.* **18**(1), 7–19.
4. Oleinick, N. L., R. L. Morris and I. Belichenko (2002) The role of apoptosis in response to photodynamic therapy: what, where, why, and how. *Photochem. Photobiol. Sci.* **1**(1), 1–21.
5. Sitnik, T. M., and B. W. Henderson (1998) The effect of fluence rate on tumor and normal tissue responses to photodynamic therapy. *Photochem. Photobiol.* **67**(4), 462–466.
6. van Geel, I. P., H. Oppelaar, J. P. Marijnissen and F. A. Stewart (1996) Influence of fractionation and fluence rate in photodynamic therapy with photofrin or mTHPC. *Radiat. Res.* **145**(5), 602–609.
7. Messmann, H., R. M. Szeimies, W. Baumler, R. Knuchel, H. Zirngibl, J. Scholmerich and A. Holstege (1997) Enhanced effectiveness of photodynamic therapy with laser light fractionation in patients with esophageal cancer. *Endoscopy* **29**(4), 275–280.
8. Curnow, A., J. C. Haller and S. G. Bown (2000) Oxygen monitoring during 5-aminolaevulinic acid induced photodynamic therapy in normal rat colon. Comparison of continuous and fractionated light regimes. *J. Photochem. Photobiol. B* **58**(2–3), 149–155.
9. Foster, T. H., R. S. Murant, R. G. Bryant, R. S. Knox, S. L. Gibson and R. Hilf (1991) Oxygen consumption and diffusion effects in photodynamic therapy. *Radiat. Res.* **126**(3), 296–303.
10. Pogue, B. W. and T. Hasan (1997) A theoretical study of light fractionation and dose-rate effects in photodynamic therapy. *Radiat. Res.* **147**(5), 551–559.
11. Henderson, B. W., S. M. Waldow, T. S. Mang, W. R. Potter, P. B. Malone and T. J. Dougherty (1985) Tumor destruction and kinetics of tumor cell death in two experimental mouse tumors following photodynamic therapy. *Cancer Res.* **45**(2), 572–576.
12. Selman, S. H., M. Kreimer-Birnbaum, J. E. Klaunig, P. J. Goldblatt, R. W. Keck and S. L. Britton (1984) Blood flow in transplantable bladder tumors treated with hematoporphyrin derivative and light. *Cancer Res.* **44**(5), 1924–1927.
13. McCord, J. M. (1987) Oxygen-derived radicals: a link between reperfusion injury and inflammation. *Fed. Proc.* **46**(7), 2402–2406.
14. Anholt, H. and J. Moan (1992) Fractionated treatment of CaD2 tumors in mice sensitized with aluminium phthlocyanine tetrasulfonate. *Cancer Lett.* **61**(3), 263–267.
15. Tsutsui, H., A. J. MacRobert, A. Curnow, A. Rogowska, G. Buonaccorsi, H. Kato and S. G. Bown (2002) Optimisation of illumination for photodynamic therapy with mTHPC on normal colon and a transplantable tumour in rats. *Lasers Med. Sci.* **17**(2), 101–109.
16. Muller, S., H. Walt, D. Dobler-Girdziunaite, D. Fiedler and U. Haller (1998) Enhanced photodynamic effects using fractionated laser light. *J. Photochem. Photobiol. B* **42**(1), 67–70.
17. Babilas, P., V. Schacht, G. Liebsch, O. S. Wolfbeis, M. Landthaler, R. M. Szeimies and C. Abels (2003) Effects of light fractionation and different fluence rates on photodynamic therapy with 5-aminolaevulinic acid in vivo. *Br. J. Cancer* **88**(9), 1462–1469.
18. Schirmer, R. H., R. L. Krauth-Siegel and G. E. Schulz (1989), Glutathione reductase. In *Coenzymes and Cofactors: Glutathione: Chemical, Biochemical, and Medical Aspect*, Vol. 3 (Edited by D. Dolphin, O. Avramoviăc and R. Poulson), pp. 553–596. Wiley, New York.
19. Plaetzer, K., T. Kiesslich, B. Krammer and P. Hammerl (2002) Characterization of the cell death modes and the associated changes in cellular energy supply in response to AlPcS4-PDT. *Photochem. Photobiol. Sci.* **1**(3), 172–177.
20. Oberdanner, C. B., T. Kiesslich, B. Krammer and K. Plaetzer (2002) Glucose is required to maintain high ATP-levels for the energy-utilizing steps during PDT-induced apoptosis. *Photochem. Photobiol.* **76**(6), 695–703.

Figure 2. ROS production after AlPcS₄-PDT (a) or hypericin-PDT (b) for fractionatedly and continuously irradiated A431 cells and/or under conditions of impaired glutathione recycling. Samples to which BCNU was added are related to the respective BCNU-containing control, whereas BCNU-free samples are presented in relation to untreated control samples (both controls without sensitizer and nonirradiated). Data represent mean values of three independent experiments ± SEM. Data were statistically compared with continuously irradiated cells (**: $P < 0.01$, *: $P < 0.05$) and to fractionatedly irradiated samples (++: $P < 0.01$, +: $P < 0.05$), respectively. A: AlPcS₄; B: hypericin; (c) continuous, (c+) continuous + BCNU, (f) fractionated, (f+) fractionated + BCNU.

Taken together, our data suggest that, with fractionated PDT, the glutathione reductase is able to recycle a significant portion of GSH during the dark intervals. This counteracts the photodynamic effect and, thus, reduces the efficiency of *in vitro* PDT. In good agreement with our results, Perotti and coworkers (37) could show that PDT on the LM2 cell line was less cytotoxic when GSH was added to the incubation medium. The scavenging effect of GSH during PDT was described as the one most important observed in this study.

However, the fractionation scheme represents an important factor because the high activity of glutathione reductase (38) could allow almost complete GSH regeneration during the dark periods of sufficient duration (e.g. 60 s, as used in the present study).

In conclusion, our results may be of interest for photodynamic-therapy applications on any well vascularized tumor/tissue in which oxygen deprivation is not to be expected. In clinical applications of photodynamic therapy, it has to be considered that GSH recycling may antagonize the therapeutic efficiency.

Photochemistry and Photobiology, 2005, 81   613

21. Mosmann, T. (1983) Rapid colorimetric assay for cellular growth and survival: application to proliferation and cytotoxicity assays. *J. Immunol. Methods* **65**(1–2), 55–63.

22. Curtin, J. F., M. Donovan and T. G. Cotter (2002) Regulation and measurement of oxidative stress in apoptosis. *J. Immunol. Methods* **265**(1–2), 49–72.

23. Royall, J. A. and H. Ischiropoulos (1993) Evaluation of 2′,7′-dichlorofluorescin and dihydrorhodamine 123 as fluorescent probes for intracellular H2O2 in cultured endothelial cells. *Arch. Biochem. Biophys.* **302**(2), 348–355.

24. Kiesslich, T., K. Plaetzer, C. B. Oberdanner, J. Berlanda, F. J. Obermair and B. Krammer (2005) Differential effects of glucose deprivation on the cellular sensitivity towards photodynamic treatment-based production of reactive oxygen species and apoptosis-induction. *FEBS Lett.* **579**(1), 185–190.

25. Messmann, H., P. Mlkvy, G. Buonaccorsi, C. L. Davies, A. J. MacRobert and S. G. Bown (1995) Enhancement of photodynamic therapy with 5-aminolaevulinic acid-induced porphyrin photosensitisation in normal rat colon by threshold and light fractionation studies. *Br. J. Cancer* **72**(3), 589–594.

26. Meister, A. and M. E. Anderson (1983) Glutathione. *Annu. Rev. Biochem.* **52**, 711–760.

27. Dickinson, D. A. and H. J. Forman (2002) Cellular glutathione and thiols metabolism. *Biochem. Pharmacol.* **64**(5–6), 1019–1026.

28. Allen, C. M., R. Langlois, W. M. Sharman, C. La Madeleine and J. E. Van Lier (2002) Photodynamic properties of amphiphilic derivatives of aluminum tetrasulfophthalocyanine. *Photochem. Photobiol.* **76**(2), 208–216.

29. Haimovici, R., T. A. Ciulla, J. W. Miller, T. Hasan, T. J. Flotte, A. G. Kenney, K. T. Schomacker and E. S. Gragoudas (2002) Localization of rose bengal, aluminum phthalocyanine tetrasulfonate, and chlorin e6 in the rabbit eye. *Retina* **22**(1), 65–74.

30. Ali, S. M. and M. Olivo (2002) Bio-distribution and subcellular localization of Hypericin and its role in PDT induced apoptosis in cancer cells. *Int. J. Oncol.* **21**(3), 531–540.

31. Agostinis, P., A. Vantieghem, W. Merlevede and P. A. de Witte (2002) Hypericin in cancer treatment: more light on the way. *Int. J. Biochem. Cell Biol.* **34**(3), 221–241.

32. Gantchev, T. G., B. J. Gowans, D. J. Hunting, J. R. Wagner and J. E. van Lier (1994) DNA strand scission and base release photosensitized by metallo-phthalocyanines. *Int. J. Radiat. Biol.* **66**(6), 705–716.

33. Ehrenberg, B., J. L. Anderson and C. S. Foote (1998) Kinetics and yield of singlet oxygen photosensitized by hypericin in organic and biological media. *Photochem. Photobiol.* **68**(2), 135–140.

34. Kamuhabwa, A. R., P. M. Agostinis, M. A. D'Hallewin, L. Baert and P. A. de Witte (2001) Cellular photodestruction induced by hypericin in AY-27 rat bladder carcinoma cells. *Photochem. Photobiol.* **74**(2), 126–132.

35. Silveira, L. R., L. Pereira Da-Silva, C. Juel and Y. Hellsten (2003) Formation of hydrogen peroxide and nitric oxide in rat skeletal muscle cells during contractions. *Free Radic. Biol. Med.* **35**(5), 455–464.

36. Myhre, O., J. M. Andersen, H. Aarnes and F. Fonnum (2003) Evaluation of the probes 2′,7′-dichlorofluorescin diacetate, luminol, and lucigenin as indicators of reactive species formation. *Biochem. Pharmacol.* **65**(10), 1575–1582.

37. Perotti, C., A. Casas and C. B. A. M. Del (2002) Scavengers protection of cells against ALA-based photodynamic therapy-induced damage. *Lasers Med. Sci.* **17**(4), 222–229.

38. Clahsen, P. C., R. M. Moison, C. A. Holtzer and H. M. Berger (1992) Recycling of glutathione during oxidative stress in erythrocytes of the newborn. *Pediatr. Res.* **32**(4), 399–402.

V. Differential effects of glucose deprivation on the cellular sensitivity towards photodynamic treatment-based production of reactive oxygen species and apoptosis induction

FEBS 29111   FEBS Letters 579 (2005) 185–190

Differential effects of glucose deprivation on the cellular sensitivity towards photodynamic treatment-based production of reactive oxygen species and apoptosis-induction

Tobias Kiesslich[1], Kristjan Plaetzer[1], Christian Benno Oberdanner[1], Juergen Berlanda[1],
Franz Josef Obermair, Barbara Krammer*

Division of Immunology and Allergology, Department of Molecular Biology, University of Salzburg, Hellbrunnerstrasse 34, 5020 Salzburg, Austria

Received 4 October 2004; revised 9 November 2004; accepted 10 November 2004

Available online 7 December 2004

Edited by Vladimir Skulachev

Abstract Photodynamic treatment (PDT) employs a photosensitizer and the light-induced formation of reactive oxygen species – antagonized by cellular antioxidant systems – for the removal of harmful cells. This study addresses the effect of altered carbohydrate metabolism on the cellular antioxidant glutathione system, and the subsequent responses to PDT. It is shown that glucose-deprivation of 18 h prior to PDT causes a reduced level of intracellular glutathione and an increased cytotoxicity of PDT. These effects can be mimicked by inhibitors of glutathione synthesis (buthionine-sulfoximine) or its regeneration (1,3-bis-(2-chlorethyl)-1-nitrosourea). Inhibited glutathione metabolism shifts the apoptotic window to lower fluences, while glucose deprivation abolishes apoptosis as a result of ATP deficiency. Our results prove evidence for manipulation of the outcome of PDT through internal metabolic pathways.
© 2004 Federation of European Biochemical Societies. Published by Elsevier B.V. All rights reserved.

Keywords: Photodynamic treatment; Reactive oxygen species; Glutathione; Pentose phosphate pathway; Glycolysis; Apoptosis

1. Introduction

Reactive oxygen species (ROS) – although being an inevitable consequence of aerobic metabolism – pose a serious threat

*Corresponding author. Fax: +43 662 8044 150.
E-mail addresses: tobias.kiesslich@sbg.ac.at (T. Kiesslich),
kristjan.plaetzer@sbg.ac.at (K. Plaetzer),
christian.oberdanner@sbg.ac.at (C.B. Oberdanner),
juergen.berlanda@sbg.ac.at (J. Berlanda),
franzjosef.obermair@sbg.ac.at (F.J. Obermair),
barbara.krammer@sbg.ac.at (B. Krammer).

[1] These authors have contributed equally to this work.

Abbreviations: AlPcS₄, aluminum (III) phthalocyanine tetrasulfonate; BCNU, 1,3-bis-(2-chloroethyl)-1-nitrosourea; BSO, DL-buthionine-sulfoximine; DCF, 2′-7′-dichlorofluorescein; DCFH-DA, 2′, 7′-dichlorodihydrofluorescein diacetate; FCS, fetal calf serum; (GF-)DMEM, (glucose-free) Dulbecco's modified Eagle's medium; GSH, reduced glutathione; GSSG, oxidized glutathione; MTT, 3-(4,5-dimethyl-2-thiazolyl)-2,5-diphenyl-2H-tetrazolium bromide; NADP(H), nicotine adenine dinucleotide phosphate (reduced); PBS, phosphate-buffered saline; PDT, photodynamic therapy; ROS, reactive oxygen species; PPP, pentose phosphate pathway

to the cellular integrity since they can oxidize almost every compound of biological origin and hence can cause severe cellular damage when produced over a certain level of quantity [1–3]. Photodynamic therapy (PDT) uses massive generation of ROS in target cells by application of photosensitizing molecules such as protoporphyrin or phthalocyanine derivatives and activation of these compounds by irradiation with light of the appropriate energy for the removal of harmful and unwanted cells [4]. By this, both modes of cell death, apoptosis and necrosis are induced in the target tissue (for review, see [5,6]).

To a certain extent, photodynamic therapy is antagonized by cellular antioxidant defense mechanisms. The antioxidative capacity of reduced glutathione (GSH) is employed to scavenge reactive oxygen intermediates yielding oxidized glutathione (GSSG), which can be regenerated to GSH by glutathione reductase under the expense of nicotine adenine dinucleotide phosphate reduced (NADPH) [7–9]. Among other carbohydrates, glucose and its metabolization in the first reactions of the pentose phosphate pathway (PPP) represent the most important source of NADPH [10,11]. From this knowledge, one might hypothesize that the availability of glucose and its metabolization in the PPP affects the cell's capability to regenerate GSH to the usual millimolar concentration range [10,12] and subsequently the cellular sensitivity towards PDT.

Besides the consequences of altered carbohydrate metabolism for the supply of NADPH, glucose deprivation (among other effects, see Section 4) also impairs cellular energetics by inhibition of glycolytic ATP production. This is of special importance for the mode of cell death induced by PDT: glycolytic ATP production may partly compensate for mitochondrial impairment during apoptosis and supply the ATP-requiring steps in active cell death [13–15]. We could prove this hypothesis in a set of experiments published recently, where glucose deprivation of photosensitized cells in vitro causes a rapid drop of ATP and subsequently an inhibition of apoptosis and a shift to necrotic cell death under conditions which normally would cause apoptosis [16]. An additional effect of impaired glycolytic ATP production may consist in reduced de novo synthesis of GSH, which requires two moles ATP per mole GSH [7,9].

By taking up these results, we expand the analysis of the effects of impaired carbohydrate metabolism to the possible influence on the antioxidative mechanisms during PDT. Employing a well established model system consisting of the

T. Kiesslich et al. / FEBS Letters 579 (2005) 185–190

hydrophilic sensitizer aluminum (III) phthalocyanine tetrasulf-onate (AlPcS$_4$) and the human epidermoid carcinoma cell line A431, the present study addresses the question whether an al-tered carbohydrate metabolism would affect the cellular ROS antioxidative mechanisms and thus the susceptibility towards PDT. The increase in sensitivity to ROS caused by conditions of glucose deprivation could be causally attributed to impaired GSH regeneration or synthesis. However, as shown in this study, artificial inhibition of GSH metabolism did not influ-ence the cell's ability to undergo apoptosis (in contrast to glu-cose-free conditions).

Aware of the fact that PDT is applied clinically in several countries for treatment of malignant and non-malignant dis-eases [4], this study proves the importance of the cellular (car-bohydrate) metabolism for influencing the cellular susceptibility towards PDT and, therefore, the efficiency of the treatment regime.

2. Materials and methods

2.1. Cell culture and photodynamic treatment

A431 human epidermoid carcinoma cells (ATCC-Nr. CRL-1555) were cultured in Dulbecco's modified Eagle's Medium (DMEM; Sigma–Aldrich, Vienna, Austria) containing 4.5 g l^{-1} glucose supple-mented with 10 mM HEPES, 4 mM L-glutamine, 1 mM Na-pyruvate, 100 U ml^{-1} penicillin, 0.1 mg ml^{-1} streptomycin and 5% fetal calf ser-um (FCS) (all from PAA-laboratories, Linz, Austria), in a humidified atmosphere at 37 °C and 7.5% CO$_2$. For measurement of caspase-3 like activity, nuclear fragmentation and ROS production, 3×10^5 cells were seeded into 30 mm petri dishes (Greiner Bio-One, Kremsmuenster, Austria); for measurement of cytotoxicity, 7500 and 12 500 cells were seeded into 96-well microplates (black walls, clear bottom; Greiner Bio-One) for samples cultured with or without glucose, respectively. Eighteen hours prior to PDT (i.e., 24 h after seeding the cells, in the log phase), the medium was replaced with 0% FCS DMEM (standard conditions, further referred to as 'DMEM') or – after a washing step with phosphate buffered saline (PBS) – with glucose-free Dulbecco's modified Eagle's medium (0% FCS, further referred to as 'GF-DMEM', Sigma–Aldrich) supplemented as above containing 10 μM AlPcS$_4$ (product no. AlPcS-834; Frontier Scientific/Porphyrin Prod-ucts, Logan, USA). When indicated, 3 mM DL-buthionine-sulfoximine (BSO; Fluka, Buchs, Switzerland) was added simultaneously with the sensitizer (18 h) and 500 μM 1,3-bis(2-chloroethyl)-1-nitrosourea (BCNU; Sigma–Aldrich) was added 1 h prior to irradiation. Immediately before irradiation, the culture supernatant was replaced by fresh 0% FCS medium (DMEM or GF-DMEM). Irradiation was performed using a red-light illumination diode array ($\lambda_{max} = 660 \pm 20$ nm, super bright diodes, product no. L53SRC-F from Kingbright Electronic, Issum, Germany) with a light power of 10 mW cm^{-2}. Sam-ples were protected from ambient light; for all experiments cells from passage number 5–15 were used.

2.2. Cytotoxicity

Metabolic/mitochondrial activity was assessed by the reduction of 3-(4,5-dimethyl-2-thiazolyl)-2,5-diphenyl-2H-tetrazolium bromide (MTT) to the insoluble blue formazan catalyzed by mitochondrial and other cellular dehydrogenases [17]. For this purpose, cells were incu-bated for 23 hours after irradiation and the medium was replaced by 100 μl of DMEM for all samples to restore the carbohydrate metabo-lism of the cells kept in glucose-free medium, which has been proven in preliminary experiments (data not shown) to allow sufficient conver-sion of MTT. After another hour, 10 μl of a solution containing MTT (5 mg ml^{-1} in PBS) was added to the microplate wells. Reduc-tion of MTT was allowed to proceed for 45 min; afterwards the med-ium was removed, the formazan dye was solubilized by addition of 100 μl 2-propanol and the absorbance was measured at 565 nm using a Spectrafluor microplate reader (Tecan, Salzburg, Austria).

2.3. Measurement of intracellular glutathione

Intracellular GSH was extracted and measured based on a protocol published by Griffith [18]. In short, cells were cultured in 60 mm petri dishes (7.5×10^5 cells/3 ml medium) under conditions described above (DMEM, BSO, BCNU, GF-DMEM; all without sensitizer). At that time where the samples are – otherwise – PDT-treated, the supernatant was removed, the cells were washed twice with PBS and harvested by trypsinization. Dilution series of the cell samples as well as of GSH (Sigma–Aldrich) were made in PBS and extracted with 1 volume 5-sul-fosalicylic acid (5%, w/v); cell debris was removed by centrifugation and the GSH released from the cells was measured by addition of 5-5'-dithio-bis(2-nitrobenzoic acid) (Ellman's reagent, absorbance at 405 nm; Spectrafluor reader, Tecan). From estimation of the cell diam-eter and cell number by means of an electronic cell counter (CASY-1, Schaerfe-Systems, Reutlingen, Germany; usually in the range of 19 μm), the intracellular concentration of GSH, $[GSH]_{ic}$, was calculated.

2.4. Measurement of intracellular ROS

The measurement of intracellular ROS is based on the oxidation of DCFH-DA (2',7'-dichlorodihydrofluorescein diacetate), which is quantified by flow cytometry [1]. In brief, 30 min prior to irradiation, cells (treated with BSO/BCNU or GF-DMEM as described above) were incubated with 50 μM DCFH-DA. Twenty minutes after irradi-ation, cells including the supernatant were harvested using Accutase (PAA-laboratories); all subsequent steps were performed on ice. The cells were spun down at $840 \times g$ and washed once with PBS; after an-other centrifugation step, the pellet was resuspended in 500 μl PBS and the resulting red fluorescing DCF (2'·7'-dichlorofluorescein) was analyzed by flow cytometry (FACS Calibur, Beckton-Dickinson, Franklin Lakes, NJ, USA). For each sample, 10 000 cells were ana-lyzed; the fluorescence signal (FL2 channel, $\lambda_{EX} = 488$ nm, $\lambda_{EM} = 580 \pm 40$ nm) was evaluated as discussed in Section 3.

2.5. Apoptosis detection (I): caspase-3 assay

As a central indicator of apoptotic cell death, 5 h post-treatment the activity of caspase-3 like proteases was analyzed by means of a fluori-genic peptide. For experimental details, see [16]. The results are related to the protein content to correct for variations in cell number assayed; finally, the data are shown as percentage of an UV-treated sample (200 mJ cm^{-2}, $\lambda_{max} = 254$ nm, Stratagene's Stratalinker, Amsterdam, Neth-erlands), which is referred to induce a homogenous apoptotic popula-tion of cells [19].

2.6. Apoptosis detection (II): nuclear fragmentation

Nuclear fragmentation during apoptosis is measured by analysis of the DNA content. This method is based on the flow cytometric analy-sis of the DNA histogram of ethanol-fixed, ribonuclease A-treated and propidium iodide (PI) stained cells, whereby cells characterized by a smaller fluorescence than the G$_1$ peak were considered as apoptotic cells ('Sub-G$_1$ peak'). For the detailed protocol, see [16]; for each sam-ple, 10 000 cells were analyzed.

2.7. Statistical evaluation

Mean values of three independent experiments ± S.E.M. or Gauss-ian Errors are shown throughout. Statistical significance was evaluated using Student's t test.

3. Results

3.1. Cellular sensitivity towards PDT and its relationship to glucose/glutathione metabolism

In order to determine the efficiency of the PDT protocol ap-plied, we employed the MTT assay which gives information on metabolic/mitochondrial activity and hereby allows an estima-tion of the overall cytotoxicity [20].

Fig. 1 shows the percentage of the MTT activity of standard PDT samples (DMEM), cells grown under glucose-free condi-tions (GF-DMEM) and those treated with inhibitors of the glutathione metabolism (BSO or BCNU) as related to the

T. Kiesslich et al. / FEBS Letters 579 (2005) 185 190   187

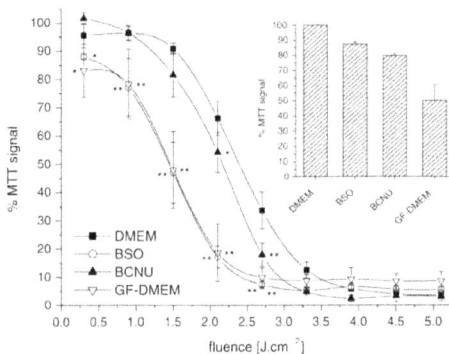

Fig. 1. Survival of photosensitized A431 cells under conditions of impaired carbohydrate/glutathione metabolism. PDT-induced cytotoxicity was assessed 24 h p.i. by means of MTT assay (Section 2.2). Results are expressed as relative MTT activity of untreated control samples; these controls (treated with BSO/BCNU/GF-DMEM but without PDT) are shown in the insertion. When indicated ('BSO'/'BCNU'), 3 mM buthionine sulfoximine was added 18 h and 500 μM 1,3-bis-(2-chloroethyl)-1-nitrosourea was added 1 h before illumination, respectively; GF-DMEM indicates glucose free medium. Mean values from three independent experiments (error bars as calculated by the Gaussian law of error propagation). Highly significant ($P < 0.01$)/significant ($P < 0.05$) differences compared to the standard PDT samples (DMEM) are indicated by ** and *, respectively.

respective controls (non-irradiated, without AlPcS$_4$, but similarly treated with GF-DMEM/BSO/BCNU). These controls showed the following MTT activities: DMEM 100%, BSO 87.3% ± 1.4, BCNU 79.8% ± 1.4 and GF-DMEM 50.1% ± 10.4 (see insertion in Fig. 1). Preceding control experiments [16] showed that this reduction is not due to cell killing by the media composition itself, but caused by reduced proliferation (data not shown). Furthermore, at a concentration of 10 μM, dark controls (data not shown) demonstrated AlPcS$_4$ not to be toxic.

All curves are characterized by a steep decrease in MTT activity with about a 2 J cm^{-2} change in light fluence ranging from 90% to 100% MTT activity to about 10% at higher light doses. Although similarly shaped, the curves for the glucose-free sample as well as those of cells with impaired GSH metabolism are clearly shifted to lower light doses as compared to the standard PDT treated cells (DMEM); a shift of about 1 J cm^{-2} is found for glucose free samples and those treated with BSO; a smaller change (approximately 0.5 J cm^{-2}) is found for BCNU treated samples.

3.2. Intracellular GSH

The intracellular concentration of GSH at the time of PDT is shown in Fig. 2. Treatment with BSO (3 mM, 18 h) reduced [GSH]$_{ic}$ from 6 mM (DMEM) to 1 mM; in contrast, the GSH level of cells treated with BCNU was not significantly reduced. The concentration of GSH in samples deprived of glucose for 18 h was nearly reduced to half (3 mM).

3.3. Generation of ROS by PDT

Determination of intracellular ROS was employed in order to clarify whether the altered cellular sensitivity caused by glucose depletion or inhibition of GSH metabolism by BSO or BCNU is caused by an alteration of ROS generation following

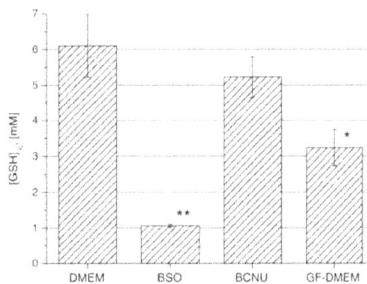

Fig. 2. Intracellular GSH under different metabolic conditions. Intracellular GSH concentration ([GSH]$_{ic}$, mM) was measured for samples with different media and additives (3 mM BSO (buthionine sulfoximine) or glucose-free medium (GF-DMEM) for 18 h and 500 μM BCNU (1,3-bis-(2-chloroethyl)-1-nitrosourea) for 1 h). Mean values of three independent measurements ± S.E.M. Highly significant ($P < 0.01$)/significant ($P < 0.05$) differences compared to the control sample (DMEM) are indicated by ** and *, respectively.

PDT. The level of intracellular ROS is changed even under control conditions (without PDT, see Fig. 3); for samples treated with GF-DMEM/BSO/BCNU but without PDT, a DCF signal for glucose-free samples and those treated with BSO or BCNU of about 58, 39, and 22 rfu (relative fluorescence units), respectively, can be found compared to control (DMEM, ~13 rfu).

For PDT, the dynamics of ROS generation in standard PDT samples (DMEM) is characterized by an increase of intracellular ROS with a small slope at low light fluences (0.25–1 J cm^{-2})

188   T. Kiesslich et al. / FEBS Letters 579 (2005) 185–190

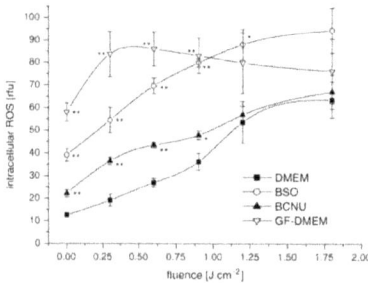

Fig. 3. Generation of intracellular ROS by photodynamic treatment. Intracellular ROS were determined by flow cytometric analysis of DCFH-DA oxidation (Section 2.4). When indicated ('BSO'/'BCNU'), 3 mM buthionine sulfoximine was added 18 h and 500 µM 1,3-bis-(2-chloroethyl)-1-nitrosourea was added 1 h before illumination, respectively; GF-DMEM indicates glucose free medium. The first data points ('0 J cm⁻²') refer to untreated controls (treated with BSO/BCNU/GF-DMEM but without PDT). Data, given as rfu (relative fluorescence units), are representative of three independent experiments ± S.E.M. Highly significant (P < 0.01)/significant (P < 0.03) differences compared to the standard PDT samples (DMEM) are indicated by ** and *, respectively.

and an increasing ROS level at higher fluences (Fig. 3). Deprivation of glucose causes a high level of DCFH-DA oxidation already at low doses, i.e., an increased level of intracellular ROS is observed within the whole range of PDT doses. Inhibition of GSH synthesis or reduction by BSO and BCNU, respectively, results in a gradual elevation of ROS generated by PDT, although the effect is more pronounced for BSO-treated samples.

At high light doses (>1.75 J cm⁻²), an additional peak appears in the DCF histogram which shows considerable lower fluorescence than the untreated control samples (about 1 order of magnitude lower; data not shown). These cells were not included in the analysis, since this peak contains most probably necrotic cells characterized by reduced membrane integrity which caused the fluorescent dye to leak into the medium.

3.4. Induction of apoptosis by PDT and its relationship to glucose/glutathione metabolism

Fig. 4A shows the activity of caspase-3 like enzymes 5 h p.i. depending on the light fluence. For standard PDT treated cells, caspase-3 activation peaked at a fluence of 2 J cm⁻² (non-irradiated controls without sensitizer are indicated by a light dose of '0' in the diagram). The maxima of caspase-3 like activity of samples treated with BSO or BCNU are clearly shifted to lower irradiation doses of about 1.5 J cm⁻². For glucose-free cells, no significant activation of capase-3 like enzymes can be found.

Results from determination of the nuclear fragmentation confirm these findings: as shown in Fig. 4B, maxima of nuclear fragmentation are present at 2 and 1.5 J cm⁻² for standard PDT samples and cells treated with BSO/BCNU, respectively; again, impairment of the GSH metabolism results in a shift of the 'apoptotic window' to lower irradiation doses. In accordance with the results from caspase activation (Fig. 4A), no

apoptotic nuclear fragmentation can be found for cells deprived of glucose.

4. Discussion

Manipulation of the source or metabolization of carbohydrates influences the pathways of cellular energy metabolism by altering the supply of substrates for glycolysis and the pentose phosphate pathway. By this, changes in the availability of glycolytic substrates will also affect the cell's ability to form NADPH. The latter represents the crucial coupling agent for a cellular mechanism involved in antioxidative defense, the glutathione system, since its regeneration (from GSSG to GSH, catalyzed by glutathione reductase) is dependent on reducing power. These facts motivated us to investigate whether the supply of glucose would influence the cellular ROS quenching mechanisms and whether an increase of the sensitivity towards ROS caused by conditions of glucose deprivation can be causally attributed to impaired GSH regeneration or synthesis. For this purpose, two different metabolic inhibitors were used to (i) deplete total glutathione by inhibition of γ-glutamyl-cystein synthase (an ATP-requiring step within the pathway of GSH de novo synthesis [1,9]; inhibited by BSO [21]) and (ii) inhibit the regeneration of GSH from the oxidized form, GSSG (catalyzed by NADPH-dependent glutathione reductase [7,9]; inhibited by BCNU [22]).

In a first set of experiments, the overall cytotoxicity of PDT under such conditions was analyzed. Both, glucose withdrawal as well as depletion of cellular glutathione (or inhibition of its regeneration) resulted in a significant increase in cytotoxicity. These results are consistent with the findings of Miller et al. [23] where addition of BSO to several cell lines caused increased cytotoxicity of Photofrin PDT and γ irradiation. The reason why BCNU resulted in a less pronounced cytotoxicity when compared to BSO may be caused by an incomplete inhibition of glutathione reductase when BCNU was applied as described; furthermore (see Fig. 2), at the time of PDT, the samples treated with BCNU are still provided with about 5 mM intracellular GSH, which is available for an initial quenching of ROS. Similar results were obtained by another group who studied the effect of glucose deprivation on the cellular sensitivity towards H₂O₂ [10]. The striking similarity between samples withdrawn of glucose as a substrate for the pentose phosphate cycle and those with hindered GSH metabolism give first evidence for a causal relationship between the increased sensitivity of glucose deprived cells and their GSH metabolism.

This suspected causal relationship of carbohydrate metabolism and the GSH system is supported by measurement of intracellular GSH (Fig. 2). Two explanations can be given for the severe reduction of [GSH]ic in glucose-free samples: (i) ROS generated under conditions of increased respiratory activity that compensates for an inhibition of glycolytic ATP production in GF-DMEM samples cause oxidation and reduction of [GSH]ic which, furthermore, cannot be regenerated due to a lack of PPP substrates; and (ii) the detrimental effect of glucose-free conditions on the cellular energy metabolism and ATP generation results in a reduced rate of the ATP-requiring de novo synthesis of GSH. It is not clear to which extent these explanations hold true for the data shown; however, both apply to the central hypothesis of this study.

J. Kiesslich et al. / FEBS Letters 579 (2005) 185–190

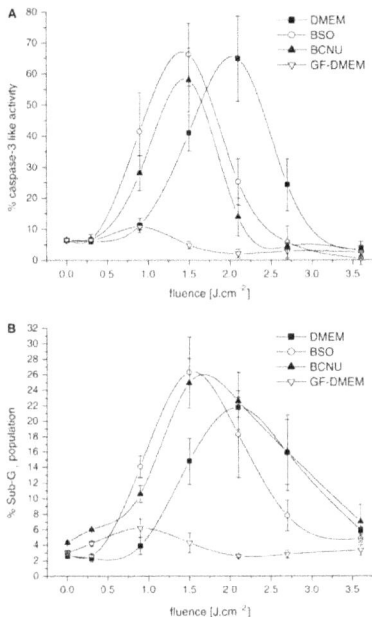

A

B

Fig. 4. Apoptosis induced by photodynamic treatment. Photosensitized cells with impaired glucose/glutathione metabolism were analyzed for caspase activation and nuclear fragmentation. When indicated ('BSO'/'BCNU'), 3 mM buthionine sulfoximine was added 18 h and 500 μM 1,3-bis-(2-chloroethyl)-1-nitrosourea was added 1 h before illumination, respectively; GF-DMEM indicates glucose free medium. (A) Activity of caspase-3 like enzymes 5 h p.i. related to UV-treated cells (a homogenous apoptotic cell population). (B) Nuclear fragmentation (Sub-G₁ peak) 8 h p.i. All data are representatives of three independent experiments ± Gaussian error bars (A) or S.E.M. (B).

The results of measurement of cytotoxicity are essentially strengthened by analysis of PDT-based generation of ROS under identical conditions. PDT applied to cells with impaired glucose or GSH metabolism results in increased intracellular ROS levels. When using DCFH-DA for measurement of ROS, it is important to note that this dye does not measure singlet oxygen itself [24], hence DCFH-DA may only give semi-quantitative information about ROS generated by PDT but yet allowing meaningful comparison within a fluence series of a given PDT treatment. The high ROS signal of GF-DMEM samples (even without PDT), therefore, could reflect other reactive oxygen species generated by cellular respiration fuelled by pyruvate [2]. Taken together, the effect of glucose deprivation/BSO or BCNU treatment on PDT-based generation of ROS is most obviously indicated by the initial increase/slope of the signal at the transition from 0 to 0.3 J cm⁻², which is – regardless of the absolute values – more pronounced for all samples compared to the standard DMEM sample. We have previously shown [16] that massive necrosis occurs under glucose-free conditions at higher light fluences; this is the reason why the ROS signal for GF-DMEM samples even decrease at higher PDT doses, since the dye is most probably released from necrotic cells.

Based on these results, we further analyzed the effect of impaired glutathione metabolism on the appearance and extent of apoptosis. Apoptosis induction could be observed for samples treated with BSO or BCNU, but the apoptotic window was shifted to lower fluences, indicating a higher susceptibility of cells with an impaired GSH system to oxidative stress. The overall extent of apoptotic cells remained unchanged. Interestingly, the shift of the apoptotic window for the BSO and BCNU samples was almost identical contrasting the different increase in cytotoxicity following PDT as shown in Fig. 1. Although BCNU is widely used as a specific inhibitor of glutathione reductase [22], this observation could be linked to the multi-functionality of this drug which may alkylate various cellular targets; thus, apart from inhibition of glutathione reductase another mode of action of BCNU might be involved in rendering cells more susceptible to induction of apoptosis.

In clear contrast, no apoptotic cells were found in samples deprived of glucose. This effect is a result of the lack of ATP in these probes and *not* of a destruction of enzymes required for apoptosis (see BSO and BCNU treated samples which clearly show apoptosis) as it was suggested by a reviewer of our earlier work [16]. For PDT, at least two studies emphasize the importance of glycolytic ATP production to compensate for a reduction of the mitochondrial membrane potential during the apoptotic process [16,25] to maintain high ATP levels after PDT [19]. Taken together, while artificial inhibition of the cell's pathways to synthesize and regenerate GSH causes an increased sensitivity towards ROS and an augmented cytotoxicity of PDT, it does not interfere with the ability to undergo active cell death. In contrast, omission of glucose not only decreases the cell's antioxidative capabilities but also abolishes the energy supply for apoptosis.

Several attempts to influence the cellular antioxidative systems have been published where: (i) systemic administration of BSO augmented phototrin-PDT mediated destruction of intracerebral 9L gliosarcoma [26], (ii) addition of several low molecular weight ROS scavengers (e.g., GSH, L-tryptophan and N-acetyl-L-cysteine) exerted a protective effect against PDT damage [27,28], and (iii) protection against PDT induced cell injury was achieved by addition of enzymes involved in the antioxidative systems (e.g., superoxide dismutase, catalase and lipoamide dehydrogenase) [27,29]. Additionally, the possibility of manipulating the cellular antioxidative metabolism by alteration of the concentration or the nature of the carbon source present in the culture medium was examined recently: substitution of glucose by galactose results in an increased rate of cell death induced by hydrogen peroxide [10,12]; an opposite effect was found for hepatocytes when fructose was added as the major carbohydrate in a hypoxia-reoxygenation model [11]. Further evidence comes from studies where: (i) addition of fructose-1,6-bisphosphate protected cortical neurons against oxidative stress partly by increasing glutathione reductase activity [30] and (ii) addition of pantothenic acid/pantothenol (precursors of coenzyme-A) increased net biosynthesis of GSH by boosting cellular energetics [31]. These studies impressively show a causal relationship between carbohydrate-,

T. Kiesslich et al. / FEBS Letters 579 (2005) 185–190

energy metabolism and the cell's antioxidative competence as expressed by the glutathione system.

It is important to note that the special case of glucose deprivation causes further consequences besides those mentioned above; among them, alteration of glucose-regulated gene expression or changes in the fluxes of metabolic pathways deriving substrates from glycolysis or the Krebs cycle may be named [32]. However, as demonstrated in this study, changes in metabolic systems which are instantly affected by glucose deprivation (such as cellular energetics and the production of reducing equivalents) and their consequences for the outcome of PDT in vitro readily can be explained without further need of taking into account specific changes in signal transduction or other more global metabolic regulation.

In conclusion, the novelty of the present study goes back to the facts that (i) it provides first evidence that management of the in vitro efficiency of PDT can be achieved not only by external administration of components involved in the antioxidative defense, but also by manipulation of the carbohydrate metabolism aiming at the control of the cellular GSH system and (ii) it also takes into account the possible impact of altered carbohydrate metabolism on the energetics and, in consequence, the mode of cell death.

References

[1] Curtin, J.F., Donovan, M. and Cotter, T.G. (2002) Regulation and measurement of oxidative stress in apoptosis. J. Immunol. Methods 265 (1–2), 49–72.

[2] Skulachev, V.P. (1999) Mitochondrial physiology and pathology: concepts of programmed death of organelles, cells and organisms. Mol. Aspects Med. 20 (3), 139–184.

[3] Skulachev, V.P. (2001) The programmed death phenomena, aging, and the Samurai law of biology. Exp. Gerontol. 36 (7), 995–1024.

[4] Dolmans, D.E., Fukumura, D. and Jain, R.K. (2003) Photodynamic therapy for cancer. Nat. Rev. Cancer 3 (5), 380–387.

[5] Oleinick, N.L., Morris, R.L. and Belichenko, I. (2002) The role of apoptosis in response to photodynamic therapy: what, where, why, and how. Photochem. Photobiol. Sci. 1 (1), 1–21.

[6] Plaetzer, K., Kiesslich, T., Verwanger, T. and Krammer, B. (2003) The modes of cell death induced by PDT: an overview. Med. Laser Appl. 18 (1), 7–19.

[7] Dickinson, D.A. and Forman, H.J. (2002) Cellular glutathione and thiols metabolism. Biochem. Pharmacol. 64 (5–6), 1019–1026.

[8] Lu, S.C. (1999) Regulation of hepatic glutathione synthesis: current concepts and controversies. FASEB J. 13 (10), 1169–1183.

[9] Meister, A. and Anderson, M.E. (1983) Glutathione. Annu. Rev. Biochem. 52, 711–760.

[10] Le Goffe, C., Vallette, G., Charrier, L., Candelon, T., Bou-Hanna, C., Bouhours, J.F. and Laboisse, C.L. (2002) Metabolic control of resistance of human epithelial cells to H2O2 and NO stresses. Biochem. J. 364 (Pt 2), 349–359.

[11] Frenzel, J., Richter, J. and Eschrich, K. (2002) Fructose inhibits apoptosis induced by reoxygenation in rat hepatocytes by decreasing reactive oxygen species via stabilization of the glutathione pool. Biochim. Biophys. Acta 1542 (1–3), 82–94.

[12] Le Goffe, C., Vallette, G., Jarry, A., Bou-Hanna, C. and Laboisse, C.L. (1999) The in vitro manipulation of carbohydrate metabolism: a new strategy for deciphering the cellular defence mechanisms against nitric oxide attack. Biochem. J. 344 (Pt 3), 643–648.

[13] Eguchi, Y., Srinivasan, A., Tomaselli, K.J., Shimizu, S. and Tsujimoto, Y. (1999) ATP-dependent steps in apoptotic signal transduction. Cancer Res. 59 (9), 2174–2181.

[14] Yasuhara, N., Eguchi, Y., Tachibana, T., Imamoto, N., Yoneda, Y. and Tsujimoto, Y. (1997) Essential role of active nuclear transport in apoptosis. Genes Cells 2 (1), 55–64.

[15] Kass, G.E., Eriksson, J.E., Weis, M., Orrenius, S. and Chow, S.C. (1996) Chromatin condensation during apoptosis requires ATP. Biochem. J. 318 (Pt 3), 749–752.

[16] Oberdanner, C.B., Kiesslich, T., Krammer, B. and Plaetzer, K. (2002) Glucose is required to maintain high ATP-levels for the energy-utilizing steps during PDT-induced apoptosis. Photochem. Photobiol. 76 (6), 695–703.

[17] Gonzalez, R.J. and Tarloff, J.B. (2001) Evaluation of hepatic subcellular fractions for Alamar blue and MTT reductase activity. Toxicol. In Vitro 15 (3), 257–259.

[18] Griffith, O.W. (1985) Glutathione and Glutathione Disulphide (Bergmeyer, H.U., Bergmeyer, J. and Grassl, M., Eds.), Methods of Enzymatic Analysis, VIII, pp. 521–529, Verlag-Chemie, Weinheim.

[19] Plaetzer, K., Kiesslich, T., Krammer, B. and Hammerl, P. (2002) Characterization of the cell death modes and the associated changes in cellular energy supply in response to AlPcS4-PDT. Photochem. Photobiol. Sci. 1 (3), 172–177.

[20] Mosmann, T. (1983) Rapid colorimetric assay for cellular growth and survival: application to proliferation and cytotoxicity assays. J. Immunol. Methods 65 (1–2), 55–63.

[21] Griffith, O.W. (1982) Mechanism of action, metabolism, and toxicity of buthionine sulfoximine and its higher homologs, potent inhibitors of glutathione synthesis. J. Biol. Chem. 257 (22), 13704–13712.

[22] Frischer, H. and Ahmad, T. (1977) Severe generalized glutathione reductase deficiency after antitumor chemotherapy with BCNU [1,3-bis(chloroethyl)-1-nitrosourea]. J. Lab. Clin. Med. 89 (5), 1080–1091.

[23] Miller, A.C. and Henderson, B.W. (1986) The influence of cellular glutathione content on cell survival following photodynamic treatment in vitro. Radiat. Res. 107 (1), 83–94.

[24] Bilski, P., Belanger, A.G. and Chignell, C.F. (2002) Photosensitized oxidation of 2′,7′-dichlorofluorescein: singlet oxygen does not contribute to the formation of fluorescent oxidation product 2′,7′-dichlorofluorescein. Free Radic. Biol. Med. 33 (7), 938–946.

[25] Kirveliene, V., Sadauskaite, A., Kadziauskas, J., Sasnauskiene, S. and Juodka, B. (2003) Correlation of death modes of photosensitized cells with intracellular ATP concentration. FEBS Lett. 553 (1–2), 167–172.

[26] Jiang, F., Lilge, L., Belcuig, M., Singh, G., Grenier, J., Li, Y. and Chopp, M. (1998) Photodynamic therapy using Photofrin in combination with buthionine sulfoximine (BSO) to treat 9L gliosarcoma in rat brain. Lasers Surg. Med. 23 (3), 161–166.

[27] Kliukiene, R., Maroziene, A., Nivinskas, H., Cenas, N., Kirveliene, V. and Juodka, B. (1997) The protective effects of dihydrolipoamide and glutathione against photodynamic damage by Alphtalocyanine tetrasulfonate. Biochem. Mol. Biol. Int. 41 (4), 707–713.

[28] Perotti, C., Casas, A. and Del, C.B.A.M. (2002) Scavengers protection of cells against ALA-based photodynamic therapy-induced damage. Lasers Med. Sci. 17 (4), 222–229.

[29] Hadjur, C., Richard, M.J., Parat, M.O., Favier, A. and Jardon, P. (1995) Photodynamically induced cytotoxicity of hypericin dye on human fibroblast cell line MRC5. J. Photochem. Photobiol. B 27 (2), 139–146.

[30] Vexler, Z.S., Wong, A., Francisco, C., Manabat, C., Christen, S., Tauber, M., Ferriero, D.M. and Gregory, G. (2003) Fructose-1,6-bisphosphate preserves intracellular glutathione and protects cortical neurons against oxidative stress. Brain Res. 960 (1–2), 90–98.

[31] Slyshenkov, V.S., Dymkowska, D. and Wojtczak, L. (2004) Pantothenic acid and pantothenol increase biosynthesis of glutathione by boosting cell energetics. FEBS Lett. 569 (1–3), 169–172.

[32] Foufelle, F., Girard, J. and Ferre, P. (1998) Glucose regulation of gene expression. Curr. Opin. Clin. Nutr. Metab. Care 1 (4), 323–328.

VI. The staurosporine analoge Ro-31-8220 induces mitochondrial apoptosis by overproduction of reactive oxygen species

The staurosporine analog Ro-31-8220 induces mitochondrial apoptosis by overproduction of reactive oxygen species

Christian Benno Oberdanner[*], Katrin Flatscher[*], Tobias Kiesslich[+], Thomas Verwanger[*], Barbara Krammer[*1], Kristjan Plaetzer[*]

[*] at:

Department of Molecular Biology, University of Salzburg, Hellbrunnerstraße 34, Salzburg, Austria

[+] at:

Department of Internal Medicine, Paracelsus Medical University and SALK, Müllner Hauptstraße 48, 5020 Salzburg, Austria

[1] to whom correspondence should be addressed at:

Ao. Prof. Dr. Barbara Krammer

Department of Molecular Biology

University of Salzburg

Hellbrunnerstraße 34

5020 Salzburg, Austria

Barbara.Krammer@sbg.ac.at

Phone: +43-662-8044-5703

Fax: +43-662-8044-150

Currently under preparation.

Keywords

Ro-31-8220, reactive oxygen species, apoptosis, mitochondria, antioxidants

Abstract

Agents with the potential to inhibit PKC have been successfully employed as effective tools in anticancer therapy because of their ability to modify cell cycle, proliferation and apoptosis. Ro-31-8220, a derivative of staurosporine (STS), is one such agent and shares with STS the ability to efficiently induce apoptosis. While STS-mediated apoptosis has been clearly linked to increased production of reactive oxygen species (ROS) inside the cell, the mechanisms of Ro-31-8220-induced apoptosis have not yet been definitely determined. In this study, we investigated the involvement of ROS in Ro-31-8220-induced apoptosis in A431 human epidermoid carcinoma cells. We therefore analyzed ROS production, cytotoxicity and selected cellular features attributed to mitochondrial apoptosis in a time-resolved manner after treatment with Ro-31-8220. We found that Ro-31-8220 application drastically increased intracellular ROS levels during the early phase of treatment and triggered a cascade of apoptotic events that ultimately led to cell death. Interestingly, Ro-31-8220-mediated ROS production was dependent on the integrity of mitochondrial membranes, indicating that the largest part of ROS is generated inside mitochondria. Moreover, ROS production and apoptosis could be significantly inhibited or even completely suppressed by pre-treatment with antioxidants. Taken together, our results provide evidence for a direct mechanistic correlation between mitochondrial ROS overproduction and apoptosis induction by Ro-31-8220.

Abbreviations

Apaf-1, apoptotic protease-activating factor-1; BCA, bicinchoninic acid; Bis, bisindolylmaleimide; c-H$_2$DCFDA, 5-(and 6-)-carboxy-2',7'-dichlorodihydrofluorescein diacetate; $\Delta\Psi$, mitochondrial membrane potential; dATP, 2'-deoxyadenosin 5'triphosphate; DEVD-AMC, 7-amino-4-methyl-coumarin; DMEM, Dulbecco's modified Eagle's medium; DMSO, dimethylsulfoxide; EDTA, ethylenediaminetetraacetate; FCS, fetal calf serum; HEPES, 2-[4-(2-hydroxyethyl)-1-piperazinyl] ethanesulfonic acid; JC-1, 5,5', 6,6'-tetrachloro-1,1', 3,3'-tetraethylbenzimidazolylcarbocyanine iodide; JNK, c-Jun N-terminal kinase; mPTP, mitochondrial permeability transition pore; MTT, 3-[4,5-dimethylthiazol-2-yl]-2,5-diphenyl tetrazolium bromide; PBS, phosphate-buffered saline; PKC, protein kinase C; ROS, reactive oxygen species; SOD, superoxide dismutase; STS, staurosporine;

Introduction

Programmed cell death – apoptosis – is an essential mechanism during the development of organisms and is thought to be conserved throughout evolution. Apoptosis plays a decisive role in various biological processes as well as several pathological phenomena and consists of a series of strictly regulated, irreversible intracellular changes [1,2].

Mitochondria play a mayor role in the apoptotic machinery, and an extensive, widespread drop of mitochondrial function is typically followed by concerted apoptotic steps , i.e. decrease of the mitochondrial membrane potential, cytochrome-c release into the cytosol, caspase activation and nuclear fragmentation [3,4].

It has become well-known that major causers for mitochondrial damage are reactive oxygen species (ROS), including highly reactive superoxide anions ($^{\bullet}O_2^-$), hydrogen peroxides (H_2O_2), which are further converted into hydroxyl radicals ($^{\bullet}OH^-$), and some other secondarily produced ROS, e.g. lipid peroxides. When the amount of ROS – either generated by the cell itself or mediated by some outer input – reaches a critical level, an opening of mitochondrial permeability transition pores (mPTPs) located in the mitochondrial membrane ensues. As a result, cytochrome-c is released from the intermembrane space into the cytosol and complexes with apoptotic protease-activating factor (Apaf)-1, pro-caspase-9 and deoxyadenosintriphosphate (dATP) to form the so-called apoptosome and, thus, initializes the apoptotic machinery [4-6].

Various environmental and chemical agents are able to induce an overproduction of ROS [7-9]. Among them, the bacterial alkaloid compound staurosporine (STS), initially characterized as an inhibitor of protein kinase C (PKC) and other kinases [10,11], has been shown to induce apoptosis in virtually all mammalian cell types [12].

One might assume that the apoptosis-inducing action of STS can be attributed to the inhibitory effect on PKC. A correlation of PKC inhibition and concomitant apoptosis induction could indeed be demonstrated in many cell types using various chemical agents [13,14]. However, since most of the studies used STS at very high concentrations that not only induce apoptosis but also non-specifically inhibit a broad range of protein kinases apart from PKC, a definite functional connection between the two processes of apoptosis induction and PKC inhibition could not be ultimately deduced [15]. Recent studies demonstrated that STS-mediated apoptosis is critically linked to an increased intracellular production of ROS, suggesting a decisive role for ROS in the induction of apoptotic processes [16-18].

A few years ago, a number of bisindolylmaleimide (Bis) compounds have been designed as analogs of STS differing only in the polyaromatic aglycon portion of the molecule. These

- 73 -

analogs were shown to be highly selective inhibitors of PKC as opposed to other protein kinases, thus providing a unique way to study the biological roles of PKC in cellular processes such as cell cycle regulation and proliferation, apoptosis, invasion, differentiation, and senescence [19-21].

Only one Bis compound, Ro-31-8220, shares with STS the ability to potently induce apoptosis, independently of the PKC inhibitory effect. Interestingly, apoptosis induction by Ro-31-8220 was shown to be inhibited by the overexpression of the apoptosis suppressor gene Bcl-2 in the model cell system, indicating that stabilization of mitochondrial pore proteins may counteract the apoptosis-inducing effect of Ro-31-8220. In this context, a thioamidine prosthetic group in the molecular structure of Ro-31-8220 was suggested to be largely responsible for the apoptosis-inducing character of the substance [22]. In addition Ro-31-8220 has been shown to induce growth arrest and apoptosis in glioblastoma cells, again with Bcl-2 being involved [23]. Other research revealed that the potential of Ro-31-8220 to activate c-Jun N-terminal kinase (JNK) might be involved in apoptosis induction by arsenic and cadmium, respectively [24,25].

However, scientific knowledge about the exact factors and mechanisms by which Ro-31-8220 induces apoptosis is still incomplete and controversial.

As a consequence, we attempted to investigate the involvement of ROS in apoptosis induced by the STS analog Ro-31-8220. We carried out a series of experiments to test the ability of Ro-31-8220 to induce cell death and heightened ROS production in human A431 epidermoid carcinoma cells. We furthermore characterized Ro-31-8220-induced cell death in regard to apoptotic features like drop of mitochondrial membrane integrity, caspase-3 activation and nuclear fragmentation. On the basis of these results, we employed several substances with distinct antioxidant properties, namely MnTBAP, Euk-8, Trolox and GSH-ee, in order to test their impact on the afore-observed Ro-31-8220-mediated effects.

Material and Methods

Cells and cell culture

A431 human epidermoid carcinoma cells (ATCC CRL-1555) were routinely cultured in standard cell culture medium (Dulbecco's modified Eagle's Medium (DMEM) with 4.5 mg.ml^{-1} glucose supplemented with 10 mM 2-[4-(2-hydroxyethyl)-piperazin-1-yl]ethanesulfonic acid (HEPES), 4 mM L-glutamine, 1 mM sodium pyruvate, 100 U.ml^{-1} penicillin, 0.1 mg.ml^{-1} streptomycin and 5% fetal calf serum (FCS), all from PAA Laboratories, Pasching, Austria) in a humidified atmosphere at 37°C and 7.5% CO_2. For all experiments, cells from passages 3-15 were used.

Treatment with Ro-31-8220 and antioxidants

Cells were incubated with 2 µM Ro-31-8220 (Sigma-Aldrich, Vienna, Austria) for varying periods (as indicated in the respective figures) in medium without FCS and in the presence or absence of antioxidants (10 µM MnTBAP (manganese (III)tctrakis(4-benzoic acid) porphyrin chloride; Calbiochem, Darmstadt, Germany), 5 µM Euk-8 (Calbiochem), 50 µM Trolox (6-hydroxy-2,5,7,8-tetramethylchroman-2-carboxylic acid; Sigma-Aldrich) and 2 mM GSH-ee (gluthatione reduced ethyl ester; Sigma-Aldrich), respectively). The optimal concentration of each antioxidant was determined in preliminary experiments (data not shown). In order to grant optimal ROS quenching, incubations with antioxidants were performed 20 h prior to Ro-31-8220 addition.

Measurement of cell viability

Cell viability was assessed by measurements of mitochondrial dehydrogenase activity via reduction of a yellow tetrazolium salt to blue formazan [26]. Cells were cultured over night and incubated with Ro-31-8220 as described above. To assess mitochondrial activity, MTT (3-(4,5-dimethylthiazol-2-yl)-2,5-diphenyltetrazolium bromide; Sigma-Aldrich) was added at a final concentration of 0.5 mg.ml^{-1} for 45 min at 37°C. The reduced formazan was dissolved in a suitable volume 2-propanol and the absorbance of triplicate samples was measured at 565 nm (Spectrafluor, Tecan, Groedig, Austria).

Measurement of intracellular ROS

Intracellular ROS were measured by the oxidation of the cell permeable, non-fluorescent dye carboxy-2′,7′dichlorodihydrofluorescein diacetate (c-H_2DCFDA, Invitrogen, Lofer, Austria) to its fluorescent form DCF, which is commonly used for the direct measurement of the redox state of cells [5,9]. Cells were treated with Ro-31-8220 for various periods and co-incubated with 100 µM c-H$_2$DCFDA 60 min prior to analysis. To those samples that were treated with Ro-31-8220 for 0, 0.5 and 1 h c-H$_2$DCFDA was added before or along with Ro-31-8220, respectively. Cells were subsequently harvested by trypsination, pelleted and washed once with phosphate buffered saline (PBS). Samples were resuspended in PBS and analyzed by flow cytometry (FL-1 channel; λ_{EX} = 488 nm, λ_{EM} = 530±30 nm) using a FACSCalibur flow cytometer (Becton Dickinson, Vienna, Austria).

Mitochondrial membrane potential

The integrity of the mitochondrial membrane potential ($\Delta\Psi$) was measured using the fluorescent probe 5,5', 6,6'-tetrachloro-1,1', 3,3'-tetraethylbenzimidazolylcarbocyanine iodide (JC-1; Sigma-Aldrich). In cells with intact membrane potential, the dye accumulates in the mitochondrial matrix and emits red fluorescence (λ_{EM} = 590 nm). Briefly, cells were incubated with 50 µg.ml^{-1} JC-1 in medium without FCS for 30 min under light-protected conditions. Cells were then harvested, washed twice with dye-free culture medium and resuspended in PBS. The red fluorescence signal of 10000 events was recorded by flow cytometry (FL-2 channel; λ_{EX} = 488 nm, λ_{EM} = 585±40 nm, FACSCalibur, Becton Dickinson) and evaluated as described previously [27].

Caspase activity

As a central indicator of apoptotic cell death, the activity of intracellular caspase-3-like proteases was assessed by measuring the specific cleavage of 7-amino-4-methylcoumarin (AMC)-labeled fluorigenic Ac-DEVD substrates (Alexis, Lausen, Switzerland) [27]. Fluorescence (λ_{EX} = 360 nm, λ_{EM} = 465 nm) was recorded on a Spectrafluor fluorimeter (Tecan) and data were corrected by the overall protein content of the cell samples measured by bicinchoninic acid (BCA) assay (Pierce, Rockford, IL, U.S.A).

Cell cycle analysis / nuclear fragmentation

Nuclear fragmentation was measured by analysis of the DNA content of cell samples. Briefly, cells were harvested, ethanol-fixed, ribonuclease A-treated and stained with propidium iodide (PI, Sigma-Aldrich). Fluorescence signals were recorded by flow cytometry (FL-2 channel; λ_{EX} = 488 nm, λ_{EM} = 585±40 nm, FACSCalibur, Becton Dickinson). *Per definitionem*, cells with a PI fluorescence below the G$_1$ peak (subG$_1$ region) are considered as apoptotic cells [27,28].

Statistical evaluation

Data represent mean of three independent experiments ± SD or SEM.

Statistical significance was calculated using Student´s t-test.

Results

Induction of cell death in A431 cells by Ro-31-8220

As a basis for further tests, a suitable Ro-31-8220 concentration was determined in a series of preliminary experiments. At a concentration of ≤ 1 μM the cells remained virtually unaffected, showing neither significant cell death nor ROS production (data not shown). Both Ro-31-8220-mediated cell death and ROS production could be ideally observed at concentrations between 2 and 3 μM. However, when higher concentrations (≥ 5 μM) were used, the cells experienced a more rapid and exhaustive demise while ROS production did not further increase (data not shown), suggesting that under these circumstances cell death is mediated by factors other than ROS production.

Based on these findings, all of the presented experiments were performed with Ro-31-8220 at a concentration of 2 μM. After various times, mitochondrial dehydrogenase activity was assessed as a common indicator for cell vitality [26]. As expected, cell viability continuously decreased with increasing incubation periods in the presence of Ro-31-8220 (Figure 1), whereas cells incubated with culture medium alone remained unaffected over the entire period (data not shown). Cell death became apparent already 2 h after Ro-31-8220 addition and reached a maximum after 18 h (<5% viable cells).

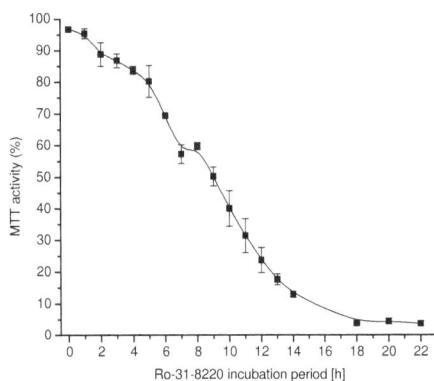

Figure 1: Effect of Ro-31-8220 on viability of A431 cells

Cells were incubated with 2 μM Ro-31-8220 for the indicated periods. Cell viability was measured by means of mitochondrial activity analyses. Data are shown as percent of untreated cells and represent the mean of three experiments ± SEM.

Production of ROS by Ro-31-8220-treated cells

Notably, the ROS concentration in Ro-31-8220-treated cells drastically increased immediately after Ro-31-8220 addition and produced a virtual 'oxidative burst' within the first 4 h of Ro-31-8220 co-incubation (Figure 2). After this period, the amount of ROS decreased to the base level and remained low over the entire period of co-incubation (data not shown). Control cells treated with medium alone showed only negligible background ROS production over the entire measurement period (approximately 30 rfu, data not shown).

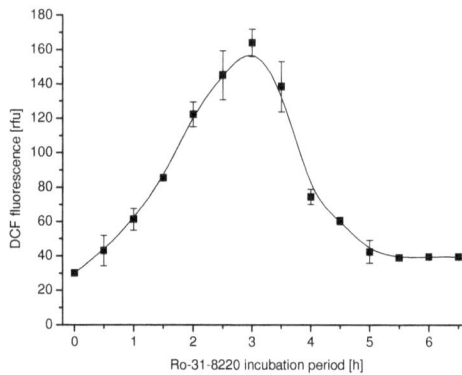

Figure 2: Ro-31-8220-induced ROS production

Cells were incubated in the presence of 2 μM Ro-31-8220 for the indicated periods. The production of intracellular ROS was visualized by staining with the fluorescent compound c-H$_2$DCFDA. The background fluorescence signal of untreated control cells stained with c-H$_2$DCFDA was approximately 30 rfu (data not shown). Results are the means ± SD of three separate experiments.

Apoptotic features of Ro-31-8220-mediated cell death

Several parameters commonly attributed to apoptotic cell death were analysed in a time-dependent manner in order to characterize Ro-31-8220-induced cell death. Since the initial steps of apoptosis are associated with a drop of mitochondrial membrane potential ($\Delta\Psi$), we applied a method detecting the membrane integrity of Ro-31-8220-treated cells. As described in a previous study [27], cells stained with JC-1 can be distinguished by their different fluorescence signals and classified as viable, apoptotic or necrotic.

Corresponding to the cell viability measurements, the fraction of the cells showing an intact mitochondrial membrane potential decreased proportionally to the Ro-31-8220 incubation period. Concurrently, the portion of cells with an "apoptotic" membrane potential increased within the same period, as did that of cells with a 'necrotic' membrane potential (Figure 3A).

As another critical marker of apoptosis we measured the activity of intracellular caspase-3-like proteases by cleavage of specific fluorigenic substrates. Figure 3B shows the activation of caspase-3 after increasing incubation periods with Ro-31-8220. Again, a distinct time course of caspase activation after Ro-31-8220 addition could be observed, starting at around 3 h of co-incubation and reaching a maximum between 9 and 13 h.

Lastly, we assessed the degree of nuclear fragmentation in Ro-31-8220-treated cells. DNA fragmentation became obvious after 4 h of Ro-31-8220 incubation and reached a maximal extent after 10-14 h (Figure 3C).

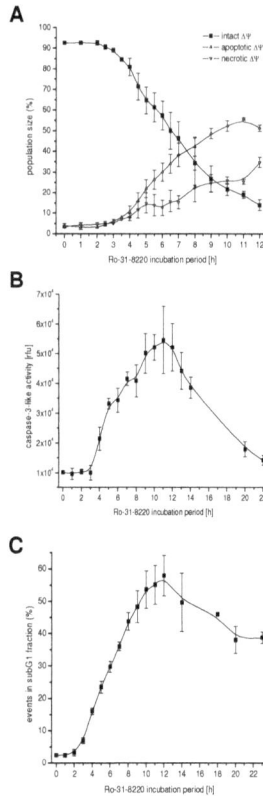

Figure 3: Induction of apoptotic features by Ro-31-8220 treatment

Cells were treated with 2 μM Ro-31-8220 for the indicated periods in the respective figures. (**A**) The mitochondrial membrane potential was measured by staining with JC-1. Cell fractions were classified as alive, apoptotic or necrotic as described in the Material and Methods section and are given as percent of the total cell population. (**B**) The activity of caspase-3-like proteases in cell samples was determined by cleavage of an AMC-labeled fluorigenic substrate. Values are expressed as fluorescence signals corrected by the overall protein content of the cell samples determined by BCA assay. (**C**) Cells were stained with propidium iodide and fluorescence was analyzed by flow cytometry. Cells with a red fluorescence below the G_1 peak were considered as apoptotic and are displayed as percent of the total cell population. All values represent the means ± SD from three independent experiments per analyzed apoptotic characteristic.

Influence of antioxidants on Ro-31-8220-induced ROS production

We employed several antioxidant agents, namely MnTBAP, Euk-8, Trolox and GSH-ee, to test their potential in quenching Ro-31-8220-mediated ROS production and, most importantly, suppressing Ro-31-8220-induced programmed cell death.

Intracellular ROS production appeared to be significantly influenced by the presence of the antioxidants. MnTBAP and Euk-8 exhibited the most potent ROS quenching capacities, whereas Trolox and GSH-ee did reduce ROS production, but to a lesser extent (Figure 4).

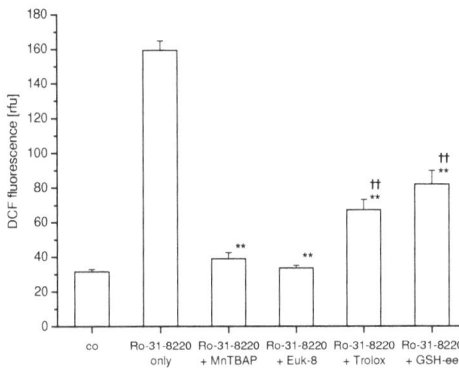

Figure 4 – Influence of antioxidants on Ro-31-8220-induced ROS production

Cells were pre-treated with the antioxidants and incubated with 2 μM Ro-31-8220 for 3h. ROS production was measured by c-H$_2$DCFDA staining as before. Data are the mean values ± SD from three independent measurements. Highly significant (P ≤ 0.01) differences compared to 'Ro-31-8220 only' are indicated with **. Highly significant signal differences of 'Trolox' and 'GSH-ee' compared to 'MnTBAP' are indicated with ††.

Influence of antioxidants on Ro-31-8220-induced cell death

As expected, each of these antioxidants was capable of suppressing Ro-31-8220-induced cell death (Figure 5). However, while MnTBAP and Euk-8 abolished the cytotoxic effect of Ro-31-8220 almost completely, i.e. >90%, Trolox and GSH-ee inhibited cell death only by 60-75%. In conformity with the cell death measurements the ROS levels were mitigated to a degree comparable to the extent of cell death attenuation measured with the respective antioxidants (compare Fig 4 and Fig 5).

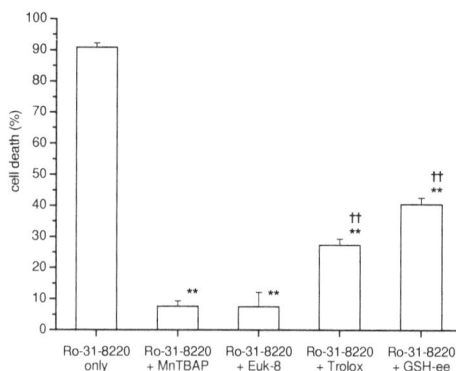

Figure 5 – Influence of antioxidants on Ro-31-8220-induced cell death

Cells were pre-incubated with the antioxidants as described in the Materials and Methods chapter and treated with 2 μM Ro-31-8220 for 20h. Cell viability was measured by means of mitochondrial activity. Results are presented as percent cell death related to untreated control cells (0% cell death). Values represent means of three experiments ± SEM. Highly significant (P ≤ 0.01) differences compared to 'Ro-31-8220 only' are indicated with **. Highly significant signal differences of 'Trolox' and 'GSH-ee' compared to 'MnTBAP' are indicated with ††.

Effects of antioxidants on apoptotic features of Ro-31-8220-mediated cell death

To optimally evaluate the influence of the different antioxidants on Ro-31-8220-induced apoptotic features, the measuring times were chosen according to those points in time when the respective features could be detected most pronouncedly in the antecedent experiments.

For example, mitochondrial membrane integrity was measured after 7 h of Ro-31-8220 treatment (Figure 6A), i.e. when maximal impact on ΔΨ was seen in the experiments without antioxidants. Accordingly, both caspase-3-like activity and DNA fragmentation were analyzed after 10 h (Figures 6B and 6C). As expected, all of these features could be significantly influenced by antioxidant treatment. Notably, the extent of attenuation of the apoptotic characteristics was comparable to the degree of ROS- and cell death-quenching observed with the respective antioxidants. Again, the most pronounced attenuation of the apoptotic features was seen with MnTBAP and Euk-8.

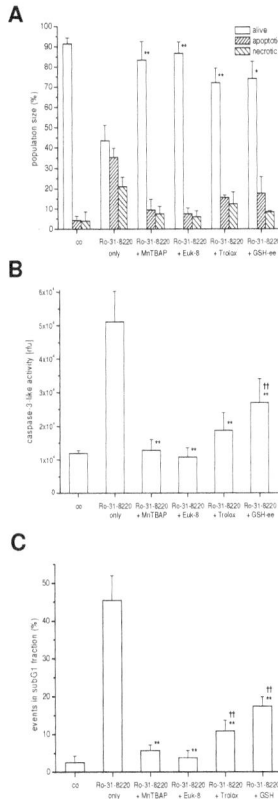

Figure 6: Influence of antioxidants on Ro-31-8220-induced apoptotic features

(A) Cells were pre-incubated with antioxidants and treated with 2 μM Ro-31-8220 for 7 h. Mitochondrial membrane integrity was assessed by JC-1 staining as before. For each sample, the percentage of cells in the alive, apoptotic and necrotic fraction, respectively, are shown.

(B) Caspase-3-like activity was measured by substrate conversion as described in cells that were pre-treated with antioxidants and incubated with Ro-31-8220 for 10 h. (C) Cells were pre-treated with antioxidants and DNA fragmentation was measured after 10 h of Ro-31-8220 treatment. In all experiments, control cells in did neither receive antioxidants nor Ro-31-8220. All values represent the means of three experiments ± SD per analyzed parameter. Highly significant ($P \leq 0.01$) and significant ($P \leq 0.05$) differences compared to 'Ro-31-8220 only' are indicated with ** and *, respectively. Highly significant signal differences of 'Trolox' and 'GSH-ee' compared to 'MnTBAP' are indicated with ††.

Discussion

In our experiments, Ro-31-8220 was shown to be capable of inducing apoptosis in the human epidermoid carcinoma cell line A431. During the early phase of Ro-31-8220 treatment, i.e. the first 3 to 4 hours, a significantly heightened production of intracellular ROS could be observed. Interestingly, after this period – corresponding to the point when cell viability started to decline – ROS production decreased again, implying that the cells are only able to generate ROS as long as their mitochondrial activity is largely unimpaired. Additional measurements revealed that the mitochondrial membrane potential was maintained over the same period as enhanced ROS production could be detected, suggesting that Ro-31-8220-induced ROS are generated inside mitochondria.

Recent findings by Santamaria *et al.* propose that staurosporine induces a heightened ROS production by acting on the mitochondrial membrane potential, precisely by causing a reverse electron flow to complex I of the mitochondrial respiratory chain [18]. This process has been shown to induce the formation of superoxide anions ($^{\bullet}O_2^-$) which – when a certain concentration is exceeded – cause the opening of mitochondrial membrane pores, ultimately leading to apoptosis induction.

In order for a sufficient amount of ROS to be produced, the integrity of the mitochondrial membrane necessarily needs to be maintained over a certain period [18,29]. However, when a critical threshold of ROS production is reached, the membrane potential drops and apoptotic mediators such as cytochrome-c and procaspase-9 are released into the cytosol.

In agreement with these findings, our results confirm a fundamental relation of the mitochondrial membrane potential to apoptosis induced by Ro-31-8220.

Hence, these data led us to conclude that Ro-31-8220 is indeed capable of inducing cell death in the carcinoma cell line A431 cells by acting on mitochondria, and that cell death is preceded by a marked increase of intracellular ROS levels. Based on these observations, we wondered whether Ro-31-8220-induced cell death is accompanied by apoptotic processes inside the cell and, therefore, analysed cellular features that are commonly attributed to apoptotic cell death [4,6]. Caspase-3 activation and nuclear fragmentation could be detected in Ro-31-8220-treated cells in a time-dependent manner. Interestingly, the apoptotic features appeared only after the ROS production was completed and the membrane potential started to decrease, i.e. after approximately 4 h.

Scheme 1 summarizes the examined changes in A431 cells after treatment with Ro-31-8220 in a time-resolved overview. It shows the time-dependent relation of ROS production, mitochondrial membrane integrity and cell viability. At a critical turning point 3-4 h after Ro-31-8220 addition ROS production reaches a maximum and, concomitantly, mitochondrial membrane potential and viability start to decrease dramatically. At the same time, the levels of caspase-3 activation and DNA fragmentation start to increase, reaching a maximum between 9 and 14 h until, after 18 h apoptosis is completed.

Taken together, apoptosis induction by Ro-31-8220 in A431 cells seems to be initiated by excess production of ROS which is followed by the characteristic apoptotic events.

We conclude that the action of Ro-31-8220-induced ROS is mainly concentrated on mitochondria and that sufficient ROS can only be produced as long as the mitochondrial membrane potential is largely intact.

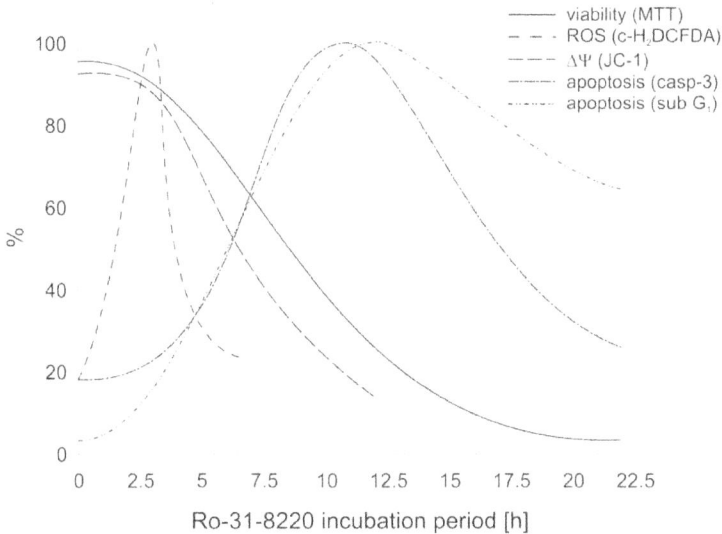

Scheme 1 –Time course of apoptosis induction by Ro-31-8220

The results from figures 1-3 are presented in a schematic overview. For convenience, the maxima of the respective experimental parameters, i.e. viability, ROS production, drop of mitochondrial membrane potential, caspase-3 activation and nuclear fragmentation, are set to 100% to show the interrelation of the functional features in a time-resolved manner. Likewise, all available data points from the individual experiments are calculated as percent of the corresponding maximal signal.

Several studies have demonstrated that compounds with antioxidant properties may exert protective effects in different situations of cellular dysfunction [30,31]. We therefore tested a series of antioxidants for their potential in preventing the observed Ro-31-8220-mediated effects.

MnTBAP is a cell permeable metalloporphyrin which mimics mitochondrial superoxide dismutase (SOD) and, in addition, has potent hydrogen peroxide scavenger properties [32,33]. Likewise, Euk-8, a synthetic selen-manganese complex, has both SOD and catalase activities [34,35].

Trolox is a water-soluble analog of alpha-tocopherol (vitamin E) with potent ROS quenching properties, especially of hydrogen peroxide. In addition, Trolox has been shown to support protein stability by inhibiting lipid peroxidation [36,37].

Gluthatione monoethyl ester (GSH-ee) is intracellularily converted to GSH which is probably the most important natural cellular antioxidant. GSH and its metabolism have been shown to be critically involved in ROS-mediated apoptotic systems [9,38-40].

We found that all four antioxidants were capable of suppressing Ro-31-8220-induced ROS production and cell death. Similarly, the apoptotic features, i.e. the drop of mitochondrial membrane potential, caspase-3-like activity and nuclear fragmentation were attenuated by the antioxidant agents. Upon closer examination, however, MnTBAP and Euk-8 exhibited the most potent inhibitory effects on all of the measured parameters and inhibited Ro-31-8220-induced apoptosis almost entirely. In contrast, Trolox and GSH-ee did suppress the Ro-31-8220-mediated effects to a certain extent, but failed to completely abolish ROS production, apoptotic events and cell death.

We attribute this to the fact that MnTBAP and Euk-8 are able to penetrate into mitochondria and may therefore immediately detoxify mitochondrial ROS at their point of origin [41,42]. In addition, MnTBAP and Euk-8 are both highly potent scavengers of superoxide anions which have been suggested to be produced by treatment with STS [18] and presumably also Ro-31-8220. Thus, MnTBAP and Euk-8 are both ideally equipped and positioned to neutralize superoxide anions produced by mitochondria. Furthermore, their catalase-like activity allows for quenching of secondary ROS like hydrogen peroxides generated during the dismutation of superoxide anions.

In contrast, Trolox and GSH-ee have been described to preferentially quench hydrogen peroxides and hydroxyl radicals which are the predominant reactive oxygen species in the cytosol. However, since both substances are hardly able to gain access to mitochondria they fail to neutralize superoxide anions, and quenching is concentrated on extramitochondrial ROS [43]. Trolox was slightly more effective than GSH-ee in suppressing the Ro-31-8220-induced effects which may be due to its protein stabilizing properties [44]. In this context, Trolox may be able to stabilize mitochondrial pore proteins that are responsible for the maintenance of mitochondrial integrity and thereby thwart mitochondrial apoptosis induction.

These findings led us to conclude that Ro-31-8220 treatment induces the production of $^\bullet O_2^-$ (maybe by acting on complex I of the mitochondrial respiratory chain). In the absence of antioxidants, these $^\bullet O_2^-$ not only act directly on mitochondrial pore complexes but also are converted into secondary ROS like H_2O_2 and $^\bullet OH^-$ [5]. Taken together, these changes inevitably result in mitochondrial disintegration and, ultimately, apoptosis. Accordingly, antioxidants that are able to immediately detoxify $^\bullet O_2^-$ inside mitochondria have the highest potential in inhibiting the Ro-31-8220-mediated effects. In contrast, compounds with antioxidant activity against secondary H_2O_2 and $^\bullet OH^-$ (like Trolox and GSH-ee) are only able to quench a lesser amount of ROS. Since MnTBAP and Euk-8 possess both SOD- and catalase-like activity, their quenching potential is most pronounced.

In general, this efficient quenching of ROS may be relevant when using Ro-31-8220 as anti-cancer tool. Many patients treated for cancer use dietary supplements with antioxidative properties, hoping to reduce the toxicity of chemotherapeutics and radiotherapy. Regarding the fact that many chemotherapeutic agents act by producing free radicals, the use of antioxidants in anti-cancer treatment is highly controversial [45,46]. If Ro-31-8220 is used because of its ability to induce apoptosis in cancer cells, one should take into account that in a certain range of concentrations the cytotoxic effect of Ro-31-8220 may be due to an overproduction of ROS, which then may be detoxified by an additional use of antioxidants.

Conclusion

Our results demonstrate that Ro-31-8220-induced apoptosis in A431 cells is mediated by mitochondrial ROS overproduction. Accordingly, treatment with antioxidants is able to suppress Ro-31-8220-induced apoptosis in dependence of ROS-specificity of the respective antioxidant and its localisation inside the cell. The better effectiveness of antioxidants with SOD- and catalase-like capacity as opposed to other ROS scavengers suggests that superoxide anions produced inside mitochondria are the major inducing factors for Ro-31-8220-mediated apoptosis and, unless neutralized, pose the most harmful threat at the cell.

However, in the light of the use of Ro-31-8220 in anti-cancer therapy, where dietary supplements, particularly antioxidants, are used to reduce the toxicity of chemotherapy, our results suggest that the desired cytotoxic effects of Ro-31-8220 may be counteracted by the antioxidative properties of vitamins and other nutritional supplements.

Our findings permit further insights into apoptosis induction and regulation by oxidative stress and provide evidence for the hypothesis of Skulachev {Skulachev, 1996 #40; Skulachev, 1999 #28; Skulachev, 2001 #20}, according to which apoptosis is tightly linked to mitochondrial ROS production, even if cell death is induced by chemical substances.

References:

1. Raff, M, (1998) Cell suicide for beginners. Nature 396: 119-22.

2. Vaux, DL and Korsmeyer, SJ, (1999) Cell death in development. Cell 96: 245-54.

3. Green, DR and Amarante-Mendes, GP, (1998) The point of no return: mitochondria, caspases, and the commitment to cell death. Results Probl Cell Differ 24: 45-61.

4. Skulachev, VP, (1999) Mitochondrial physiology and pathology; concepts of programmed death of organelles, cells and organisms. Mol Aspects Med 20: 139-84.

5. Curtin, JF, Donovan, M and Cotter, TG, (2002) Regulation and measurement of oxidative stress in apoptosis. J Immunol Methods 265: 49-72.

6. Skulachev, VP, (2001) The programmed death phenomena, aging, and the Samurai law of biology. Exp Gerontol 36: 995-1024.

7. Bresgen, N, Karlhuber, G, Krizbai, I, Bauer, H, Bauer, HC and Eckl, PM, (2003) Oxidative stress in cultured cerebral endothelial cells induces chromosomal aberrations, micronuclei, and apoptosis. J Neurosci Res 72: 327-33.

8. Liu, BH, Wu, TS, Yu, FY and Su, CC, (2007) Induction of oxidative stress response by the mycotoxin patulin in mammalian cells. Toxicol Sci 95: 340-7.

9. Oberdanner, CB, Plaetzer, K, Kiesslich, T and Krammer, B, (2005) Photodynamic treatment with fractionated light decreases production of reactive oxygen species and cytotoxicity in vitro via regeneration of glutathione. Photochem Photobiol 81: 609-13.

10. Herbert, JM, Seban, E and Maffrand, JP, (1990) Characterization of specific binding sites for [3H]-staurosporine on various protein kinases. Biochem Biophys Res Commun 171: 189-95.

11. Ruegg, UT and Burgess, GM, (1989) Staurosporine, K-252 and UCN-01: potent but nonspecific inhibitors of protein kinases. Trends Pharmacol Sci 10: 218-20.

12. Bertrand, R, Solary, E, O'Connor, P, Kohn, KW and Pommier, Y, (1994) Induction of a common pathway of apoptosis by staurosporine. Exp Cell Res 211: 314-21.

13. Jarvis, WD, Turner, AJ, Povirk, LF, Traylor, RS and Grant, S, (1994) Induction of apoptotic DNA fragmentation and cell death in HL-60 human promyelocytic leukemia cells by pharmacological inhibitors of protein kinase C. Cancer Res 54: 1707-14.

14. Lucas, M and Sanchez-Margalet, V, (1995) Protein kinase C involvement in apoptosis. Gen Pharmacol 26: 881-7.

15. Harkin, ST, Cohen, GM and Gescher, A, (1998) Modulation of apoptosis in rat thymocytes by analogs of staurosporine: lack of direct association with inhibition of protein kinase C. Mol Pharmacol 54: 663-70.

16. Gil, J, Almeida, S, Oliveira, CR and Rego, AC, (2003) Cytosolic and mitochondrial ROS in staurosporine-induced retinal cell apoptosis. Free Radic Biol Med 35: 1500-14.

17. Pong, K, Doctrow, SR, Huffman, K, Adinolfi, CA and Baudry, M, (2001) Attenuation of staurosporine-induced apoptosis, oxidative stress, and mitochondrial dysfunction by synthetic superoxide dismutase and catalase mimetics, in cultured cortical neurons. Exp Neurol 171: 84-97.

18. Santamaria, G, Martinez-Diez, M, Fabregat, I and Cuezva, JM, (2006) Efficient execution of cell death in non-glycolytic cells requires the generation of ROS controlled by the activity of mitochondrial H+-ATP synthase. Carcinogenesis 27: 925-35.

19. Davis, PD, Elliott, LH, Harris, W, Hill, CH, Hurst, SA, Keech, E et al., (1992) Inhibitors of protein kinase C. 2. Substituted bisindolylmaleimides with improved potency and selectivity. J Med Chem 35: 994-1001.

20. Davis, PD, Hill, CH, Lawton, G, Nixon, JS, Wilkinson, SE, Hurst, SA et al., (1992) Inhibitors of protein kinase C. 1. 2,3-Bisarylmaleimides. J Med Chem 35: 177-84.

21. Serova, M, Ghoul, A, Benhadji, KA, Cvitkovic, E, Faivre, S, Calvo, F et al., (2006) Preclinical and clinical development of novel agents that target the protein kinase C family. Semin Oncol 33: 466-78.

22. Han, Z, Pantazis, P, Lange, TS, Wyche, JH and Hendrickson, EA, (2000) The staurosporine analog, Ro-31-8220, induces apoptosis independently of its ability to inhibit protein kinase C. Cell Death Differ 7: 521-30.

23. Begemann, M, Kashimawo, SA, Lunn, RM, Delohery, T, Choi, YJ, Kim, S et al., (1998) Growth inhibition induced by Ro 31-8220 and calphostin C in human glioblastoma cell lines is associated with apoptosis and inhibition of CDC2 kinase. Anticancer Res 18: 3139-52.

24. Qu, W, Bortner, CD, Sakurai, T, Hobson, MJ and Waalkes, MP, (2002) Acquisition of apoptotic resistance in arsenic-induced malignant transformation: role of the JNK signal transduction pathway. Carcinogenesis 23: 151-9.

25. Qu, W, Fuquay, R, Sakurai, T and Waalkes, MP, (2006) Acquisition of apoptotic resistance in cadmium-induced malignant transformation: specific perturbation of JNK signal transduction pathway and associated metallothionein overexpression. Mol Carcinog 45: 561-71.

26.   Mosmann, T, (1983) Rapid colorimetric assay for cellular growth and survival: application to proliferation and cytotoxicity assays. J Immunol Methods 65: 55-63.

27.   Oberdanner, CB, Kiesslich, T, Krammer, B and Plaetzer, K, (2002) Glucose is required to maintain high ATP-levels for the energy-utilizing steps during PDT-induced apoptosis. Photochem Photobiol 76: 695-703.

28.   Ormerod, MG, (2002) Investigating the relationship between the cell cycle and apoptosis using flow cytometry. J Immunol Methods 265: 73-80.

29.   Votyakova, TV and Reynolds, IJ, (2001) DeltaPsi(m)-Dependent and -independent production of reactive oxygen species by rat brain mitochondria. J Neurochem 79: 266-77.

30.   Hermans, N, Cos, P, Maes, L, De Bruyne, T, Vanden Berghe, D, Vlietinck, AJ et al., (2007) Challenges and pitfalls in antioxidant research. Curr Med Chem 14: 417-30.

31.   Seifried, HE, Anderson, DE, Fisher, EI and Milner, JA, (2007) A review of the interaction among dietary antioxidants and reactive oxygen species. J Nutr Biochem.

32.   Cuzzocrea, S, Costantino, G, Mazzon, E, Zingarelli, B, De Sarro, A and Caputi, AP, (1999) Protective effects of Mn(III)tetrakis (4-benzoic acid) porphyrin (MnTBAP), a superoxide dismutase mimetic, in paw oedema induced by carrageenan in the rat. Biochem Pharmacol 58: 171-6.

33.   Day, BJ, Fridovich, I and Crapo, JD, (1997) Manganic porphyrins possess catalase activity and protect endothelial cells against hydrogen peroxide-mediated injury. Arch Biochem Biophys 347: 256-62.

34.   McDonald, MC, d'Emmanuele di Villa Bianca, R, Wayman, NS, Pinto, A, Sharpe, MA, Cuzzocrea, S et al., (2003) A superoxide dismutase mimetic with catalase activity (EUK-8) reduces the organ injury in endotoxic shock. Eur J Pharmacol 466: 181-9.

35.   van Empel, VP, Bertrand, AT, van Oort, RJ, van der Nagel, R, Engelen, M, van Rijen, HV et al., (2006) EUK-8, a superoxide dismutase and catalase mimetic, reduces cardiac oxidative stress and ameliorates pressure overload-induced heart failure in the harlequin mouse mutant. J Am Coll Cardiol 48: 824-32.

36.   Cerecetto, H and Lopez, GV, (2007) Antioxidants derived from vitamin E: an overview. Mini Rev Med Chem 7: 315-38.

37.   Wu, TW, Hashimoto, N, Wu, J, Carey, D, Li, RK, Mickle, DA et al., (1990) The cytoprotective effect of Trolox demonstrated with three types of human cells. Biochem Cell Biol 68: 1189-94.

38.   Anderson, ME, Powrie, F, Puri, RN and Meister, A, (1985) Glutathione monoethyl ester: preparation, uptake by tissues, and conversion to glutathione. Arch Biochem Biophys 239: 538-48.

39.   Estrela, JM, Ortega, A and Obrador, E, (2006) Glutathione in cancer biology and therapy. Crit Rev Clin Lab Sci 43: 143-81.

40.   Kiesslich, T, Plaetzer, K, Oberdanner, CB, Berlanda, J, Obermair, FJ and Krammer, B, (2005) Differential effects of glucose deprivation on the cellular sensitivity towards photodynamic treatment-based production of reactive oxygen species and apoptosis-induction. FEBS Lett 579: 185-90.

41.   Hinerfeld, D, Traini, MD, Weinberger, RP, Cochran, B, Doctrow, SR, Harry, J et al., (2004) Endogenous mitochondrial oxidative stress: neurodegeneration, proteomic analysis, specific respiratory chain defects, and efficacious antioxidant therapy in superoxide dismutase 2 null mice. J Neurochem 88: 657-67.

42.   Szabo, C, Day, BJ and Salzman, AL, (1996) Evaluation of the relative contribution of nitric oxide and peroxynitrite to the suppression of mitochondrial respiration in immunostimulated macrophages using a manganese mesoporphyrin superoxide dismutase mimetic and peroxynitrite scavenger. FEBS Lett 381: 82-6.

43.   Szeto, HH, (2006) Mitochondria-targeted peptide antioxidants: novel neuroprotective agents. Aaps J 8: E521-31.

44.   Dean, RT, Hunt, JV, Grant, AJ, Yamamoto, Y and Niki, E, (1991) Free radical damage to proteins: the influence of the relative localization of radical generation, antioxidants, and target proteins. Free Radic Biol Med 11: 161-8.

45.   D'Andrea, GM, (2005) Use of antioxidants during chemotherapy and radiotherapy should be avoided. CA Cancer J Clin 55: 319-21.

46.   Seely, D, Stempak, D and Baruchel, S, (2007) A strategy for controlling potential interactions between natural health products and chemotherapy: a review in pediatric oncology. J Pediatr Hematol Oncol 29: 32-47.

VII. References of Introduction

1. Skulachev, V.P., *Role of uncoupled and non-coupled oxidations in maintenance of safely low levels of oxygen and its one-electron reductants.* Q Rev Biophys, 1996. **29**(2): p. 169-202.

2. Lissi, E.A., et al., *Singlet oxygen O2(1.DELTA.g) bimolecular processes. Solvent and compartmentalization effects.* Chem Rev, 1993. **93**(2): p. 699-723.

3. Halliwell, B.a.G., *Free Radicals in Biology and Medicine.* Vol. 4. 2007, New York: Oxford University Press Inc.

4. Skulachev, V.P., *Mitochondrial physiology and pathology; concepts of programmed death of organelles, cells and organisms.* Mol Aspects Med, 1999. **20**(3): p. 139-84.

5. Papa, S. and V.P. Skulachev, *Reactive oxygen species, mitochondria, apoptosis and aging.* Mol Cell Biochem, 1997. **174**(1-2): p. 305-19.

6. Fenton, H.J.H., *The Oxidation of Polyhydric Alcohols in Presence of Iron. .* Journal of Chemical Society, 1894. **75**: p. 1-11.

7. Haber, F.W., J.J., *The catlytic decomposition of hydrogen peroxide by iron salts.* Proc. R. Soc. Lond. Ser. A, 1934. **147**: p. 332-352.

8. Curtin, J.F., M. Donovan, and T.G. Cotter, *Regulation and measurement of oxidative stress in apoptosis.* J Immunol Methods, 2002. **265**(1-2): p. 49-72.

9. Andreyev, A.Y., Y.E. Kushnareva, and A.A. Starkov, *Mitochondrial metabolism of reactive oxygen species.* Biochemistry (Mosc), 2005. **70**(2): p. 200-14.

10. Tolkien, J.R.R., *The lord of the rings*, ed. G.A. Unwin. 1945.

11. Whatley, S.A., et al., *Superoxide, neuroleptics and the ubiquinone and cytochrome b5 reductases in brain and lymphocytes from normals and schizophrenic patients.* Mol Psychiatry, 1998. **3**(3): p. 227-37.

12. Kunduzova, O.R., et al., *Hydrogen peroxide production by monoamine oxidase during ischemia/reperfusion.* Eur J Pharmacol, 2002. **448**(2-3): p. 225-30.

13. Maurel, A., et al., *Age-dependent increase in hydrogen peroxide production by cardiac monoamine oxidase A in rats.* Am J Physiol Heart Circ Physiol, 2003. **284**(4): p. H1460-7.

14. Carvalho, F., et al., *Hydrogen peroxide production in mouse tissues after acute d-amphetamine administration. Influence of monoamine oxidase inhibition.* Arch Toxicol, 2001. **75**(8): p. 465-9.

15. Kumar, M.J., D.G. Nicholls, and J.K. Andersen, *Oxidative alpha-ketoglutarate dehydrogenase inhibition via subtle elevations in monoamine oxidase B levels results in*

loss of spare respiratory capacity: implications for Parkinson's disease. J Biol Chem, 2003. **278**(47): p. 46432-9.

16. Loffler, M., et al., *Catalytic enzyme histochemistry and biochemical analysis of dihydroorotate dehydrogenase/oxidase and succinate dehydrogenase in mammalian tissues, cells and mitochondria.* Histochem Cell Biol, 1996. **105**(2): p. 119-28.

17. Kwong, L.K. and R.S. Sohal, *Substrate and site specificity of hydrogen peroxide generation in mouse mitochondria.* Arch Biochem Biophys, 1998. **350**(1): p. 118-26.

18. Zhang, L., L. Yu, and C.A. Yu, *Generation of superoxide anion by succinate-cytochrome c reductase from bovine heart mitochondria.* J Biol Chem, 1998. **273**(51): p. 33972-6.

19. Vasquez-Vivar, J., B. Kalyanaraman, and M.C. Kennedy, *Mitochondrial aconitase is a source of hydroxyl radical. An electron spin resonance investigation.* J Biol Chem, 2000. **275**(19): p. 14064-9.

20. Tretter, L. and V. Adam-Vizi, *Generation of reactive oxygen species in the reaction catalyzed by alpha-ketoglutarate dehydrogenase.* J Neurosci, 2004. **24**(36): p. 7771-8.

21. Starkov, A.A., et al., *Mitochondrial alpha-ketoglutarate dehydrogenase complex generates reactive oxygen species.* J Neurosci, 2004. **24**(36): p. 7779-88.

22. Grigolava, I.V., et al., *[Tiron as a spin-trap for superoxide radicals produced by the respiratory chain of submitochondrial particles].* Biokhimiia, 1980. **45**(1): p. 75-82.

23. Loschen, G. and A. Azzi, *Proceedings: Formation of oxygen radicals and hydrogen peroxide in mitochondrial membranes.* Hoppe Seylers Z Physiol Chem, 1974. **355**(10): p. 1226.

24. Loschen, G. and A. Azzi, *On the formation of hydrogen peroxide and oxygen radicals in heart mitochondria.* Recent Adv Stud Cardiac Struct Metab, 1975. **7**: p. 3-12.

25. Boveris, A., E. Cadenas, and A.O. Stoppani, *Role of ubiquinone in the mitochondrial generation of hydrogen peroxide.* Biochem J, 1976. **156**(2): p. 435-44.

26. Turrens, J.F., *Mitochondrial formation of reactive oxygen species.* J Physiol, 2003. **552**(Pt 2): p. 335-44.

27. Turrens, J.F., *Superoxide production by the mitochondrial respiratory chain.* Biosci Rep, 1997. **17**(1): p. 3-8.

28. Skulachev, V.P., *Membrane Bioenergetics*, ed. N.Y. Springer Verlag. 1988.

29. Liu, Y., G. Fiskum, and D. Schubert, *Generation of reactive oxygen species by the mitochondrial electron transport chain.* J Neurochem, 2002. **80**(5): p. 780-7.

30. Korshunov, S.S., V.P. Skulachev, and A.A. Starkov, *High protonic potential actuates a mechanism of production of reactive oxygen species in mitochondria.* FEBS Lett, 1997. **416**(1): p. 15-8.

31.   Korshunov, S.S., et al., *Fatty acids as natural uncouplers preventing generation of O2.-and H2O2 by mitochondria in the resting state.* FEBS Lett, 1998. **435**(2-3): p. 215-8.

32.   Starkov, A.A. and G. Fiskum, *Regulation of brain mitochondrial H2O2 production by membrane potential and NAD(P)H redox state.* J Neurochem, 2003. **86**(5): p. 1101-7.

33.   Hansford, R.G., B.A. Hogue, and V. Mildaziene, *Dependence of H2O2 formation by rat heart mitochondria on substrate availability and donor age.* J Bioenerg Biomembr, 1997. **29**(1): p. 89-95.

34.   Loschen, G., L. Flohe, and B. Chance, *Respiratory chain linked H(2)O(2) production in pigeon heart mitochondria.* FEBS Lett, 1971. **18**(2): p. 261-264.

35.   Starkov, A.A., B.M. Polster, and G. Fiskum, *Regulation of hydrogen peroxide production by brain mitochondria by calcium and Bax.* J Neurochem, 2002. **83**(1): p. 220-8.

36.   Kushnareva, Y., A.N. Murphy, and A. Andreyev, *Complex I-mediated reactive oxygen species generation: modulation by cytochrome c and NAD(P)+ oxidation-reduction state.* Biochem J, 2002. **368**(Pt 2): p. 545-53.

37.   Votyakova, T.V. and I.J. Reynolds, *DeltaPsi(m)-Dependent and -independent production of reactive oxygen species by rat brain mitochondria.* J Neurochem, 2001. **79**(2): p. 266-77.

38.   Skulachev, V.P., *Mitochondria in the programmed death phenomena; a principle of biology: "it is better to die than to be wrong".* IUBMB Life, 2000. **49**(5): p. 365-73.

39.   Hatch, G.M., D.E. Vance, and D.C. Wilton, *Rat liver mitochondrial phospholipase A2 is an endotoxin-stimulated membrane-associated enzyme of Kupffer cells which is released during liver perfusion.* Biochem J, 1993. **293** (**Pt 1**): p. 143-50.

40.   Flint, D.H., et al., *The role and properties of the iron-sulfur cluster in Escherichia coli dihydroxy-acid dehydratase.* J Biol Chem, 1993. **268**(20): p. 14732-42.

41.   Gardner, P.R., et al., *Superoxide radical and iron modulate aconitase activity in mammalian cells.* J Biol Chem, 1995. **270**(22): p. 13399-405.

42.   Martin, D.S., D. Spriggs, and J.A. Koutcher, *A concomitant ATP-depleting strategy markedly enhances anticancer agent activity.* Apoptosis, 2001. **6**(1-2): p. 125-31.

43.   Ott, M., et al., *Mitochondria, oxidative stress and cell death.* Apoptosis, 2007. **12**(5): p. 913-22.

44.   Elstner, E.F., R.J. Youngman, and W. Oßwald, *Superoxide dismutase*, in *Methods of enzymatic analysis, Enzymes 1*, H.U. Bergmeyer, J. Bergmeyer, and M. Grassl, Editors. 1985, VCH: Weinheim. p. 273-282.

45.   Li, Y., et al., *Dilated cardiomyopathy and neonatal lethality in mutant mice lacking manganese superoxide dismutase.* Nat Genet, 1995. **11**(4): p. 376-81.

46.   Lebovitz, R.M., et al., *Neurodegeneration, myocardial injury, and perinatal death in mitochondrial superoxide dismutase-deficient mice.* Proc Natl Acad Sci U S A, 1996. **93**(18): p. 9782-7.

47.   Aebi, H.E., *Catalase,* in *Methods of enzymatic analysis, Enzymes 1,* H.U. Bergmeyer, J. Bergmeyer, and M. Grassl, Editors. 1985, VCH: Weinheim. p. 273-282.

48.   Lledias, F., P. Rangel, and W. Hansberg, *Oxidation of catalase by singlet oxygen.* J Biol Chem, 1998. **273**(17): p. 10630-7.

49.   Bai, J., et al., *Overexpression of catalase in cytosolic or mitochondrial compartment protects HepG2 cells against oxidative injury.* J Biol Chem, 1999. **274**(37): p. 26217-24.

50.   Tome, M.E., et al., *Catalase-overexpressing thymocytes are resistant to glucocorticoid-induced apoptosis and exhibit increased net tumor growth.* Cancer Res, 2001. **61**(6): p. 2766-73.

51.   Ho, Y.S., et al., *Mice lacking catalase develop normally but show differential sensitivity to oxidant tissue injury.* J Biol Chem, 2004. **279**(31): p. 32804-12.

52.   Antunes, F., D. Han, and E. Cadenas, *Relative contributions of heart mitochondria glutathione peroxidase and catalase to $H(2)O(2)$ detoxification in in vivo conditions.* Free Radic Biol Med, 2002. **33**(9): p. 1260-7.

53.   Skulachev, V.P., *Cytochrome c in the apoptotic and antioxidant cascades.* FEBS Lett, 1998. **423**(3): p. 275-80.

54.   Tomasetti, M., et al., *Coenzyme Q10 enrichment decreases oxidative DNA damage in human lymphocytes.* Free Radic Biol Med, 1999. **27**(9-10): p. 1027-32.

55.   Sakano, K., et al., *Suppression of azoxymethane-induced colonic premalignant lesion formation by coenzyme Q10 in rats.* Asian Pac J Cancer Prev, 2006. **7**(4): p. 599-603.

56.   Cup&Tracy, *Coenzyme Q10,* in *Dietary Supplements,* T. Human Press, New Jersy, Editor. 2003.

57.   Linnane, A.W., *Cellular coenzyme Q10 redox poise constitutes a major cell metabolic and gene regulatory system.* Biogerontology, 2002. **3**(1-2): p. 3-6.

58.   Linnane, A.W., et al., *Cellular redox activity of coenzyme Q10: effect of CoQ10 supplementation on human skeletal muscle.* Free Radic Res, 2002. **36**(4): p. 445-53.

59.   LeWitt, P.A., *Neuroprotection for Parkinson's disease.* J Neural Transm Suppl, 2006(71): p. 113-22.

60.   Jacob, R.A. and B.J. Burri, *Oxidative damage and defense.* Am J Clin Nutr, 1996. **63**(6): p. 985S-990S.

61.   Niki, E., *Interaction of ascorbate and alpha-tocopherol.* Ann N Y Acad Sci, 1987. **498**: p. 186-99.

62.   Packer, L., *Oxidants, antioxidant nutrients and the athlete.* J Sports Sci, 1997. **15**(3): p. 353-63.

63.   Packer, L., *Molecular mechanisms and protective effects of vitamin C in atherosclerosis.* Journal of Nutrition, 2005. **131**: p. 369-373.

64.   Woodall, A.A., et al., *Oxidation of carotenoids by free radicals: relationship between structure and reactivity.* Biochim Biophys Acta, 1997. **1336**(1): p. 33-42.

65.   Sergio, A.R., *Beat-carotene and other carotenoids as antioxidants.* Journal of the American College of Nutrition, 1999. **18**: p. 426-433.

66.   Susin, S.A., N. Zamzami, and G. Kroemer, *Mitochondria as regulators of apoptosis: doubt no more.* Biochim Biophys Acta, 1998. **1366**(1-2): p. 151-65.

67.   Yang, J., et al., *Prevention of apoptosis by Bcl-2: release of cytochrome c from mitochondria blocked.* Science, 1997. **275**(5303): p. 1129-32.

68.   Green, D.R., *Apoptotic pathways: the roads to ruin.* Cell, 1998. **94**(6): p. 695-8.

69.   Mirkovic, N., et al., *Resistance to radiation-induced apoptosis in Bcl-2-expressing cells is reversed by depleting cellular thiols.* Oncogene, 1997. **15**(12): p. 1461-70.

70.   McCullough, K.D., et al., *Gadd153 sensitizes cells to endoplasmic reticulum stress by down-regulating Bcl2 and perturbing the cellular redox state.* Mol Cell Biol, 2001. **21**(4): p. 1249-59.

71.   Hockenbery, D.M., et al., *Bcl-2 functions in an antioxidant pathway to prevent apoptosis.* Cell, 1993. **75**(2): p. 241-51.

72.   Kane, D.J., et al., *Bcl-2 inhibition of neural death: decreased generation of reactive oxygen species.* Science, 1993. **262**(5137): p. 1274-7.

73.   Skulachev, V.P., *NAD(P)(+) decomposition and antioxidant defense of the cell.* FEBS Lett, 2001. **492**(1-2): p. 1-3.

74.   Tischler, M.E., P. Hecht, and J.R. Williamson, *Effect of ammonia on mitochondrial and cytosolic NADH and NADPH systems in isolated rat liver cells.* FEBS Lett, 1977. **76**(1): p. 99-104.

75.   Meister, A. and M.E. Anderson, *Glutathione.* Annu Rev Biochem, 1983. **52**: p. 711-60.

76.   Schirmer, R.H., R.L. Krauth-Siegel, and G.E. Schulz, *Glutathione reductase*, in *Coenzymes and cofactors; Glutathione : chemical, biochemical, and medical aspects*, D. Dolphin, O. Avramoviâc, and R. Poulson, Editors. 1989, Wiley: New York. p. 553-596.

77.   Maiorino, M., et al., *Reactivity of phospholipid hydroperoxide glutathione peroxidase with membrane and lipoprotein lipid hydroperoxides.* Free Radic Res Commun, 1991. **12-13 Pt 1**: p. 131-5.

78. Thomas, J.P., et al., *Protective action of phospholipid hydroperoxide glutathione peroxidase against membrane-damaging lipid peroxidation. In situ reduction of phospholipid and cholesterol hydroperoxides.* J Biol Chem, 1990. **265**(1): p. 454-61.

79. Imai, H. and Y. Nakagawa, *Biological significance of phospholipid hydroperoxide glutathione peroxidase (PHGPx, GPx4) in mammalian cells.* Free Radic Biol Med, 2003. **34**(2): p. 145-69.

80. Dringen, R., *Metabolism and functions of glutathione in brain.* Prog Neurobiol, 2000. **62**(6): p. 649-71.

81. Griffith, O.W., *Biologic and pharmacologic regulation of mammalian glutathione synthesis.* Free Radic Biol Med, 1999. **27**(9-10): p. 922-35.

82. Bremer, H.J., *Disturbance of Amino Acid Metabolism*, in *Clinicla Chemistry and Dagnosis*, B.-M. Urban and Schwarzenberger, Editor. 1981. p. 80-82.

83. Griffith, O.W. and A. Meister, *Potent and specific inhibition of glutathione synthesis by buthionine sulfoximine (S-n-butyl homocysteine sulfoximine).* J Biol Chem, 1979. **254**(16): p. 7558-60.

84. Frischer, H. and T. Ahmad, *Severe generalized glutathione reductase deficiency after antitumor chemotherapy with BCNU'' [1,3-bis(chloroethyl)-1-nitrosourea].* J Lab Clin Med, 1977. **89**(5): p. 1080-91.

85. Oda, T., et al., *Specific efflux of glutathione from the basolateral membrane domain in polarized MDCK cells during ricin-induced apoptosis.* J Biochem (Tokyo), 1999. **126**(4): p. 715-21.

86. Kroemer, G., B. Dallaporta, and M. Resche-Rigon, *The mitochondrial death/life regulator in apoptosis and necrosis.* Annu Rev Physiol, 1998. **60**: p. 619-42.

87. Wullner, U., et al., *Glutathione depletion potentiates MPTP and MPP+ toxicity in nigral dopaminergic neurones.* Neuroreport, 1996. **7**(4): p. 921-3.

88. Jiang, F., et al., *Photodynamic therapy using Photofrin in combination with buthionine sulfoximine (BSO) to treat 9L gliosarcoma in rat brain.* Lasers Surg Med, 1998. **23**(3): p. 161-6.

89. Kliukiene, R., et al., *The protective effects of dihydrolipoamide and glutathione against photodynamic damage by Al-phtalocyanine tetrasulfonate.* Biochem Mol Biol Int, 1997. **41**(4): p. 707-13.

90. Tan, S., et al., *The regulation of reactive oxygen species production during programmed cell death.* J Cell Biol, 1998. **141**(6): p. 1423-32.

91. Gul, M., et al., *Cellular and clinical implications of glutathione.* Indian J Exp Biol, 2000. **38**(7): p. 625-34.

92. Estrela, J.M., A. Ortega, and E. Obrador, *Glutathione in cancer biology and therapy.* Crit Rev Clin Lab Sci, 2006. **43**(2): p. 143-81.

93. Miller, A.C. and B.W. Henderson, *The influence of cellular glutathione content on cell survival following photodynamic treatment in vitro.* Radiat Res, 1986. **107**(1): p. 83-94.

94. Weber, G.F., *Final common pathways in neurodegenerative diseases: regulatory role of the glutathione cycle.* Neurosci Biobehav Rev, 1999. **23**(8): p. 1079-86.

95. Cuzzocrea, S., et al., *Protective effects of Mn(III)tetrakis (4-benzoic acid) porphyrin (MnTBAP), a superoxide dismutase mimetic, in paw oedema induced by carrageenan in the rat.* Biochem Pharmacol, 1999. **58**(1): p. 171-6.

96. Cuzzocrea, S., et al., *Beneficial effects of Mn(III)tetrakis (4-benzoic acid) porphyrin (MnTBAP), a superoxide dismutase mimetic, in carrageenan-induced pleurisy.* Free Radic Biol Med, 1999. **26**(1-2): p. 25-33.

97. Day, B.J., I. Fridovich, and J.D. Crapo, *Manganic porphyrins possess catalase activity and protect endothelial cells against hydrogen peroxide-mediated injury.* Arch Biochem Biophys, 1997. **347**(2): p. 256-62.

98. McDonald, M.C., et al., *A superoxide dismutase mimetic with catalase activity (EUK-8) reduces the organ injury in endotoxic shock.* Eur J Pharmacol, 2003. **466**(1-2): p. 181-9.

99. van Empel, V.P., et al., *EUK-8, a superoxide dismutase and catalase mimetic, reduces cardiac oxidative stress and ameliorates pressure overload-induced heart failure in the harlequin mouse mutant.* J Am Coll Cardiol, 2006. **48**(4): p. 824-32.

100. Cerecetto, H. and G.V. Lopez, *Antioxidants derived from vitamin E: an overview.* Mini Rev Med Chem, 2007. **7**(3): p. 315-38.

101. Wu, T.W., et al., *The cytoprotective effect of Trolox demonstrated with three types of human cells.* Biochem Cell Biol, 1990. **68**(10): p. 1189-94.

102. Anderson, M.E., et al., *Glutathione monoethyl ester: preparation, uptake by tissues, and conversion to glutathione.* Arch Biochem Biophys, 1985. **239**(2): p. 538-48.

103. Skulachev, V.P., *[Decrease in the intracellular concentration of O2 as a special function of the cellular respiratory system].* Biokhimiia, 1994. **59**(12): p. 1910-2.

104. Skulachev, V.P., *Why are mitochondria involved in apoptosis? Permeability transition pores and apoptosis as selective mechanisms to eliminate superoxide-producing mitochondria and cell.* FEBS Lett, 1996. **397**(1): p. 7-10.

105. Skulachev, V.P., *The programmed death phenomena, aging, and the Samurai law of biology.* Exp Gerontol, 2001. **36**(7): p. 995-1024.

106. Skulachev, V.P., *Uncoupling: new approaches to an old problem of bioenergetics.* Biochim Biophys Acta, 1998. **1363**(2): p. 100-24.

107. Lawen, A., *Apoptosis-an introduction.* Bioessays, 2003. **25**(9): p. 888-96.

108. Green, D.R. and G.P. Amarante-Mendes, *The point of no return: mitochondria, caspases, and the commitment to cell death.* Results Probl Cell Differ, 1998. **24**: p. 45-61.

109.   Nicholson, D.W., *Caspase structure, proteolytic substrates, and function during apoptotic cell death.* Cell Death Differ, 1999. **6**(11): p. 1028-42.

110.   Creagh, E.M. and S.J. Martin, *Cell stress-associated caspase activation: intrinsically complex?* Sci STKE, 2003. **2003**(175): p. pe11.

111.   Villa, P., S.H. Kaufmann, and W.C. Earnshaw, *Caspases and caspase inhibitors.* Trends Biochem Sci, 1997. **22**(10): p. 388-93.

112.   Hengartner, M.O., *The biochemistry of apoptosis.* Nature, 2000. **407**(6805): p. 770-6.

113.   Liu, X., et al., *DFF, a heterodimeric protein that functions downstream of caspase-3 to trigger DNA fragmentation during apoptosis.* Cell, 1997. **89**(2): p. 175-84.

114.   Sakahira, H., M. Enari, and S. Nagata, *Cleavage of CAD inhibitor in CAD activation and DNA degradation during apoptosis.* Nature, 1998. **391**(6662): p. 96-9.

115.   Adams, J.M., *Ways of dying: multiple pathways to apoptosis.* Genes Dev, 2003. **17**(20): p. 2481-95.

116.   Green, D.R. and J.C. Reed, *Mitochondria and apoptosis.* Science, 1998. **281**(5381): p. 1309-12.

117.   Schwartzman, R.A. and J.A. Cidlowski, *Apoptosis: the biochemistry and molecular biology of programmed cell death.* Endocr Rev, 1993. **14**(2): p. 133-51.

118.   Scaffidi, C., et al., *Two CD95 (APO-1/Fas) signaling pathways.* Embo J, 1998. **17**(6): p. 1675-87.

119.   Grutter, M.G., *Caspases: key players in programmed cell death.* Curr Opin Struct Biol, 2000. **10**(6): p. 649-655.

120.   Modjtahedi, N., et al., *Apoptosis-inducing factor: vital and lethal.* Trends Cell Biol, 2006. **16**(5): p. 264-72.

121.   Zhang, J., et al., *Endonuclease G is required for early embryogenesis and normal apoptosis in mice.* Proc Natl Acad Sci U S A, 2003. **100**(26): p. 15782-7.

122.   Bajt, M.L., et al., *Nuclear translocation of endonuclease G and apoptosis-inducing factor during acetaminophen-induced liver cell injury.* Toxicol Sci, 2006. **94**(1): p. 217 25.

123.   Albright, C.D., et al., *Mitochondrial and microsomal derived reactive oxygen species mediate apoptosis induced by transforming growth factor-beta1 in immortalized rat hepatocytes.* J Cell Biochem, 2003. **89**(2): p. 254-61.

124.   Slater, A.F., C.S. Nobel, and S. Orrenius, *The role of intracellular oxidants in apoptosis.* Biochim Biophys Acta, 1995. **1271**(1): p. 59-62.

125.   Sugiyama, H., et al., *Reactive oxygen species induce apoptosis in cultured human mesangial cells.* J Am Soc Nephrol, 1996. **7**(11): p. 2357-63.

126.   Albrecht, H., J. Tschopp, and C.V. Jongeneel, *Bcl-2 protects from oxidative damage and apoptotic cell death without interfering with activation of NF-kappa B by TNF.* FEBS Lett, 1994. **351**(1): p. 45-8.

127.   Wang, J.H., et al., *Induction of human endothelial cell apoptosis requires both heat shock and oxidative stress responses.* Am J Physiol, 1997. **272**(5 Pt 1): p. C1543-51.

128.   Li, G.X., et al., *Differential involvement of reactive oxygen species in apoptosis induced by two classes of selenium compounds in human prostate cancer cells.* Int J Cancer, 2007. **120**(9): p. 2034-43.

129.   Ricci, J.E., R.A. Gottlieb, and D.R. Green, *Caspase-mediated loss of mitochondrial function and generation of reactive oxygen species during apoptosis.* J Cell Biol, 2003. **160**(1): p. 65-75.

130.   Crompton, M., *The mitochondrial permeability transition pore and its role in cell death.* Biochem J, 1999. **341 (Pt 2)**: p. 233-49.

131.   Crompton, M., et al., *The mitochondrial permeability transition pore.* Biochem Soc Symp, 1999. **66**: p. 167-79.

132.   Papadopoulos, V., *Peripheral-type benzodiazepine/diazepam binding inhibitor receptor: biological role in steroidogenic cell function.* Endocr Rev, 1993. **14**(2): p. 222-40.

133.   Decaudin, D., *Peripheral benzodiazepine receptor and its clinical targeting.* Anticancer Drugs, 2004. **15**(8): p. 737-45.

134.   Tatton, W.G. and R.M. Chalmers-Redman, *Mitochondria in neurodegenerative apoptosis: an opportunity for therapy?* Ann Neurol, 1998. **44**(3 Suppl 1): p. S134-41.

135.   Desagher, S. and J.C. Martinou, *Mitochondria as the central control point of apoptosis.* Trends Cell Biol, 2000. **10**(9): p. 369-77.

136.   Crompton, M., H. Ellinger, and A. Costi, *Inhibition by cyclosporin A of a Ca2+-dependent pore in heart mitochondria activated by inorganic phosphate and oxidative stress.* Biochem J, 1988. **255**(1): p. 357-60.

137.   Chinopoulos, C., A.A. Starkov, and G. Fiskum, *Cyclosporin A-insensitive permeability transition in brain mitochondria: inhibition by 2-aminoethoxydiphenyl borate.* J Biol Chem, 2003. **278**(30): p. 27382-9.

138.   Halestrap, A.P., K.Y. Woodfield, and C.P. Connern, *Oxidative stress, thiol reagents, and membrane potential modulate the mitochondrial permeability transition by affecting nucleotide binding to the adenine nucleotide translocase.* J Biol Chem, 1997. **272**(6): p. 3346-54.

139.   Kowaltowski, A.J. and R.F. Castilho, *Ca2+ acting at the external side of the inner mitochondrial membrane can stimulate mitochondrial permeability transition induced by phenylarsine oxide.* Biochim Biophys Acta, 1997. **1322**(2-3): p. 221-9.

140.   Marzo, I., et al., *The permeability transition pore complex: a target for apoptosis regulation by caspases and bcl-2-related proteins.* J Exp Med, 1998. **187**(8): p. 1261-71.

141.   Schendel, S.L., M. Montal, and J.C. Reed, *Bcl-2 family proteins as ion-channels*. Cell Death Differ, 1998. **5**(5): p. 372-80.

142.   Marzo, I., et al., *Bax and adenine nucleotide translocator cooperate in the mitochondrial control of apoptosis*. Science, 1998. **281**(5385): p. 2027-31.

143.   De Giorgi, F., et al., *The permeability transition pore signals apoptosis by directing Bax translocation and multimerization*. Faseb J, 2002. **16**(6): p. 607-9.

144.   Tsujimoto, Y., T. Nakagawa, and S. Shimizu, *Mitochondrial membrane permeability transition and cell death*. Biochim Biophys Acta, 2006. **1757**(9-10): p. 1297-300.

145.   Borner, C., *The Bcl-2 protein family: sensors and checkpoints for life-or-death decisions*. Mol Immunol, 2003. **39**(11): p. 615-47.

146.   Zoratti, M. and I. Szabo, *The mitochondrial permeability transition*. Biochim Biophys Acta, 1995. **1241**(2): p. 139-76.

147.   Petronilli, V., C. Cola, and P. Bernardi, *Modulation of the mitochondrial cyclosporin A-sensitive permeability transition pore. II. The minimal requirements for pore induction underscore a key role for transmembrane electrical potential, matrix pH, and matrix Ca2+*. J Biol Chem, 1993. **268**(2): p. 1011-6.

148.   Bernardi, P., *Modulation of the mitochondrial cyclosporin A-sensitive permeability transition pore by the proton electrochemical gradient. Evidence that the pore can be opened by membrane depolarization*. J Biol Chem, 1992. **267**(13): p. 8834-9.

149.   Susin, S.A., et al., *The central executioner of apoptosis: multiple connections between protease activation and mitochondria in Fas/APO-1/CD95- and ceramide-induced apoptosis*. J Exp Med, 1997. **186**(1): p. 25-37.

150.   Gil, J., et al., *Cytosolic and mitochondrial ROS in staurosporine-induced retinal cell apoptosis*. Free Radic Biol Med, 2003. **35**(11): p. 1500-14.

151.   Berlanda, J., et al., *Characterization of apoptosis induced by photodynamic treatment with hypericin in A431 human epidermoid carcinoma cells*. J Environ Pathol Toxicol Oncol, 2006. **25**(1-2): p. 173-88.

152.   Oberdanner, C.B., et al., *Glucose is required to maintain high ATP-levels for the energy-utilizing steps during PDT-induced apoptosis*. Photochem Photobiol, 2002. **76**(6): p. 695-703.

153.   Costantini, P., et al., *Oxidation of a critical thiol residue of the adenine nucleotide translocator enforces Bcl-2-independent permeability transition pore opening and apoptosis*. Oncogene, 2000. **19**(2): p. 307-14.

154.   Huser, J., C.E. Rechenmacher, and L.A. Blatter, *Imaging the permeability pore transition in single mitochondria*. Biophys J, 1998. **74**(4): p. 2129-37.

155.   Kim, J.S., Y. Jin, and J.J. Lemasters, *Reactive oxygen species, but not Ca2+ overloading, trigger pH- and mitochondrial permeability transition-dependent death of adult rat*

myocytes after ischemia-reperfusion. Am J Physiol Heart Circ Physiol, 2006. **290**(5): p. H2024-34.

156.   He, L. and J.J. Lemasters, *Regulated and unregulated mitochondrial permeability transition pores: a new paradigm of pore structure and function?* FEBS Lett, 2002. **512**(1-3): p. 1-7.

157.   Bernardi, P., *Mitochondrial transport of cations: channels, exchangers, and permeability transition.* Physiol Rev, 1999. **79**(4): p. 1127-55.

158.   Lam, M., N.L. Oleinick, and A.L. Nieminen, *Photodynamic therapy-induced apoptosis in epidermoid carcinoma cells. Reactive oxygen species and mitochondrial inner membrane permeabilization.* J Biol Chem, 2001. **276**(50): p. 47379-86.

159.   Piper, H.M., et al., *Cytosolic Ca2+ overload and macromolecule permeability of endothelial monolayers.* Herz, 1992. **17**(5): p. 277-83.

160.   Grebenova, D., et al., *Mitochondrial and endoplasmic reticulum stress-induced apoptotic pathways are activated by 5-aminolevulinic acid-based photodynamic therapy in HL60 leukemia cells.* J Photochem Photobiol B, 2003. **69**(2): p. 71-85.

161.   Chakraborti, T., et al., *Oxidant, mitochondria and calcium: an overview.* Cell Signal, 1999. **11**(2): p. 77-85.

162.   Kowaltowski, A.J., R.F. Castilho, and A.E. Vercesi, *Ca(2+)-induced mitochondrial membrane permeabilization: role of coenzyme Q redox state.* Am J Physiol, 1995. **269**(1 Pt 1): p. C141-7.

163.   Kowaltowski, A.J., R.F. Castilho, and A.E. Vercesi, *Mitochondrial permeability transition and oxidative stress.* FEBS Lett, 2001. **495**(1-2): p. 12-5.

164.   Usuda, J., et al., *Association between the photodynamic loss of Bcl-2 and the sensitivity to apoptosis caused by phthalocyanine photodynamic therapy.* Photochem Photobiol, 2003. **78**(1): p. 1-8.

165.   Oleinick, N.L., R.L. Morris, and I. Belichenko, *The role of apoptosis in response to photodynamic therapy: what, where, why, and how.* Photochem Photobiol Sci, 2002. **1**(1): p. 1-21.

166.   Rizzuto, R., G. Pitton, and G.F. Azzone, *Effect of Ca2+, peroxides, SH reagents, phosphate and aging on the permeability of mitochondrial membranes.* Eur J Biochem, 1987. **162**(2): p. 239-49.

167.   Crompton, M. and A. Costi, *A heart mitochondrial Ca2(+)-dependent pore of possible relevance to re-perfusion-induced injury. Evidence that ADP facilitates pore interconversion between the closed and open states.* Biochem J, 1990. **266**(1): p. 33-9.

168.   Novgorodov, S.A., et al., *Ion permeability induction by the SH cross-linking reagents in rat liver mitochondria is inhibited by the free radical scavenger, butylhydroxytoluene.* J Bioenerg Biomembr, 1987. **19**(3): p. 191-202.

169. Costantini, P., R. Colonna, and P. Bernardi, *Induction of the mitochondrial permeability transition by N-ethylmaleimide depends on secondary oxidation of critical thiol groups. Potentiation by copper-ortho-phenanthroline without dimerization of the adenine nucleotide translocase.* Biochim Biophys Acta, 1998. **1365**(3): p. 385-92.

170. Iwata, S., et al., *Adult T cell leukemia (ATL)-derived factor/human thioredoxin prevents apoptosis of lymphoid cells induced by L-cystine and glutathione depletion: possible involvement of thiol-mediated redox regulation in apoptosis caused by pro-oxidant state.* J Immunol, 1997. **158**(7): p. 3108-17.

171. Skulachev, V.P., *Phenoptosis: programmed death of an organism.* Biochemistry (Mosc), 1999. **64**(12): p. 1418-26.

172. Steinmetz, K.A. and J.D. Potter, *Vegetables, fruit, and cancer prevention: a review.* J Am Diet Assoc, 1996. **96**(10): p. 1027-39.

173. Seifried, H.E., et al., *The antioxidant conundrum in cancer.* Cancer Res, 2003. **63**(15): p. 4295-8.

174. Willett, W.C. and D. Trichopoulos, *Nutrition and cancer: a summary of the evidence.* Cancer Causes Control, 1996. **7**(1): p. 178-80.

175. Blot, W.J., et al., *Nutrition intervention trials in Linxian, China: supplementation with specific vitamin/mineral combinations, cancer incidence, and disease-specific mortality in the general population.* J Natl Cancer Inst, 1993. **85**(18): p. 1483-92.

176. Hwang, H., J. Dwyer, and R.M. Russell, *Diet, Helicobacter pylori infection, food preservation and gastric cancer risk: are there new roles for preventative factors?* Nutr Rev, 1994. **52**(3): p. 75-83.

177. Bostick, R.M., et al., *Reduced risk of colon cancer with high intake of vitamin E: the Iowa Women's Health Study.* Cancer Res, 1993. **53**(18): p. 4230-7.

178. Heinonen, O.P., et al., *Prostate cancer and supplementation with alpha-tocopherol and beta-carotene: incidence and mortality in a controlled trial.* J Natl Cancer Inst, 1998. **90**(6): p. 440-6.

179. Cooke, M.S., et al., *Oxidative DNA damage: mechanisms, mutation, and disease.* Faseb J, 2003. **17**(10): p. 1195-214.

180. Salganik, R.I., *The benefits and hazards of antioxidants: controlling apoptosis and other protective mechanisms in cancer patients and the human population.* J Am Coll Nutr, 2001. **20**(5 Suppl): p. 464S-472S; discussion 473S-475S.

181. Omenn, G.S., et al., *Effects of a combination of beta carotene and vitamin A on lung cancer and cardiovascular disease.* N Engl J Med, 1996. **334**(18): p. 1150-5.

182. Salganik, R.I., et al., *Dietary antioxidant depletion: enhancement of tumor apoptosis and inhibition of brain tumor growth in transgenic mice.* Carcinogenesis, 2000. **21**(5): p. 909-14.

183. Byers, T. and G. Perry, *Dietary carotenes, vitamin C, and vitamin E as protective antioxidants in human cancers.* Annu Rev Nutr, 1992. **12**: p. 139-59.

184. Miyajima, A., et al., *Role of reactive oxygen species in cis-dichlorodiammineplatinum-induced cytotoxicity on bladder cancer cells.* Br J Cancer, 1997. **76**(2): p. 206-10.

185. Jarvis, W.D., et al., *Induction of apoptotic DNA fragmentation and cell death in HL-60 human promyelocytic leukemia cells by pharmacological inhibitors of protein kinase C.* Cancer Res, 1994. **54**(7): p. 1707-14.

186. Zhang, X.D., S.K. Gillespie, and P. Hersey, *Staurosporine induces apoptosis of melanoma by both caspase-dependent and -independent apoptotic pathways.* Mol Cancer Ther, 2004. **3**(2): p. 187-97.

187. Bertrand, R., et al., *Induction of a common pathway of apoptosis by staurosporine.* Exp Cell Res, 1994. **211**(2): p. 314-21.

188. Lucas, M. and V. Sanchez-Margalet, *Protein kinase C involvement in apoptosis.* Gen Pharmacol, 1995. **26**(5): p. 881-7.

189. Harkin, S.T., G.M. Cohen, and A. Gescher, *Modulation of apoptosis in rat thymocytes by analogs of staurosporine: lack of direct association with inhibition of protein kinase C.* Mol Pharmacol, 1998. **54**(4): p. 663-70.

190. Santamaria, G., et al., *Efficient execution of cell death in non-glycolytic cells requires the generation of ROS controlled by the activity of mitochondrial H+-ATP synthase.* Carcinogenesis, 2006. **27**(5): p. 925-35.

191. Pong, K., et al., *Attenuation of staurosporine-induced apoptosis, oxidative stress, and mitochondrial dysfunction by synthetic superoxide dismutase and catalase mimetics, in cultured cortical neurons.* Exp Neurol, 2001. **171**(1): p. 84-97.

192. Perotti, C., A. Casas, and C.B.A.M. Del, *Scavengers protection of cells against ALA-based photodynamic therapy-induced damage.* Lasers Med Sci, 2002. **17**(4): p. 222-9.

VIII. Appendix

(i) **Review by Pleatzer et. al. 2005:**

"Apoptosis Following Photodynamic Therapy: Induction, Mechanism and Detection"

Current Pharmaceutical Design, 2005, *11*, 1151-1165
1151

Apoptosis Following Photodynamic Tumor Therapy: Induction, Mechanisms and Detection

Kristjan Plaetzer, Tobias Kiesslich, Christian Benno Oberdanner and Barbara Krammer*

Department of Molecular Biology, University of Salzburg, Hellbrunnerstrasse 34, 5020 Salzburg, Austria

Abstract: As a treatment modality for malign and certain non-malignant diseases, photodynamic therapy (PDT) involves a two step protocol which consists of the (selective) uptake and accumulation of a photosensitizing agent in target cells and the subsequent irradiation with light in the visible range. Reactive oxygen species (ROS) produced during this process cause cellular damage and, depending on the treatment dose / severity of damage, lead to either cellular repair / survival, apoptotic cell death or necrosis. PDT-induced apoptosis has been focused on during the last years due to the intimate connection between ROS generation, mitochondria and apoptosis; by this PDT employs mechanisms different to those in the action of radio- and chemotherapeutics, giving rise to the chance of apoptosis induction by PDT even in cells resistant to conventional treatments. In this review, the (experimental) variables determining the cellular response after PDT and the known mechanistic details of PDT-triggered induction and execution of apoptosis are discussed. This is accompanied by a critical evaluation of wide-spread methods employed in apoptosis detection with special respect to *in vitro* / cell-based methodology.

Key Words: Photodynamic therapy, anticancer treatment, apoptosis, cell death, apoptosis detection methods.

1. INTRODUCTION

Reactive oxygen species (ROS) represent a cellular stress factor that can – if produced over a certain level of quantity – effectively induce the active mode of cell death, apoptosis [1-3]. This sensitivity of cells towards ROS overproduction can be used for the removal of harmful or unwanted cells. Photodynamic therapy (PDT) utilizes this effect to treat several types of early stage tumors and other diseases [4-7]. Typically, PDT involves a two step protocol; firstly, a light-absorbing compound (the photosensitizer) is taken up (mostly) specifically by target cells [8-12], secondly, the sensitizer is activated by irradiation with visible light at the appropriate energy [13-15]. Since the light penetration depth in tissue increases with the wavelength, an absorption maximum at the highest possible wavelength (red / near infrared) is taken for activation of the sensitizer [16-20].

Upon absorption of a photon the photosensitizer is energized into a high-energy singlet state, from which it may change into the triplet state by intersystem crossing. The triplet state of these molecules is relatively long-lasting ($\geq 10^{-3}$ seconds [21]) and can exert the transfer of an electron to adjacent molecules (preferably oxygen; referred to as a type I photochemical reaction) or energy to molecular oxygen (type II photochemical reaction) [13]. The transfer of an electron to molecular oxygen produces the superoxide anion, which can form hydrogen peroxide H_2O_2. Due to its ability to diffuse through membranes, the latter might be toxic for neighbouring cells as well. By taking up another electron, H_2O_2 can split up into two hydroxyl radicals – the most dangerous member of the ROS family ($E_0 = +1.35V$) –

which can attack and oxidize any compound of biological origin; additionally, these processes are facilitated by low activation energies [22-24]. In simple chemical systems, especially if porphyrins are used as photosensitizers, PDT has been shown to predominantly exert its effect by type II photochemical reaction. The latter generates singlet oxygen, a highly reactive molecule, which reacts with many biomolecules [25-28]. The operational sequence of PDT is outlined in Fig. (1).

The overall response of cells to PDT depends on external parameters, such as the incubation protocol or the light dose applied and on internal parameters, subsumed under 'cellular *susceptibility*' towards PDT. The outcome of such treatment can be classified as either (i) repair and survival of target cells, (ii) apoptosis or active cell death or (iii) necrosis or passive cell death, the latter causing leakage of cell content into tissue and subsequent inflammatory processes (reviewed in [29]). Since PDT has been shown to be a potent inducer of apoptosis [26, 30-33] and in some cases can trigger active cell death even in cells, which have turned out to be unable to undergo apoptosis after chemo- or radiotherapy [34], this review shall focus on the mechanisms, which lead from the excited photosensitizer to the induction and execution of apoptosis. The last chapter shall deal with detection methods for apoptosis and – at least to a certain level – weight against their respective benefits and drawbacks.

1.1. Active Cell Death: Key Factors, Mechanisms and Implications on Cells and Tissues

The increasing interest in apoptosis is based on two facts. Firstly, in many cases, apoptosis can be found at lower doses of medical treatment than those required to elicit necrosis [35, 36]. The so-called 'apoptotic window' (we shall use this term in the following as an indicative of a certain range of treatment doses which causes apoptosis) can be found 'in

*Address correspondence to this author at the Department of Molecular Biology, University of Salzburg, Hellbrunnerstrasse 34, 5020 Salzburg, Austria; Tel: +43 662 8044 5703; Fax: +43 662 8044 150; E-mail: Barbara.Krammer@sbg.ac.at

1381-6128/05 $50.00+.00

1152   *Current Pharmaceutical Design, 2005, Vol. 11, No. 9*   *Plaetzer et al.*

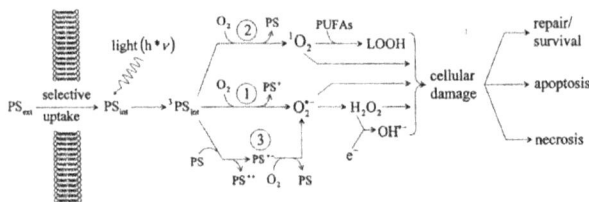

Fig. (1). Schematic representation of the principle of photodynamic treatment. PDT involves the (selective) uptake of a photosensitizing agent (PS) from the extracellular space (PS_{ext}) into target cells (PS_{int}) and the subsequent activation by irradiation with visible light ($h^\bullet v$). In the following, the activated sensitizer ($^3PS_{int}$) can either (1) transfer an electron to a nearby target molecule amid formation of the superoxide anion (ROS, type I photochemical reaction) or (2) energy to adjacent molecules yielding singlet oxygen (type II photochemical reaction) or (3) activate surrounding photosensitizing molecules. Superoxide anions may be transformed into the cell permeant H_2O_2, which - in turn - can split up into hydroxyl radicals. Singlet oxygen can induce a chain reaction of lipids creating long-living peroxyl radicals and thus lead to an amplification of ROS production and subsequent photodynamic damage. Activated photosensitizing molecules form superoxide radicals as well. The cellular damage induced by these reactive species may be repaired or lead to cell death via apoptosis or necrosis.

between' survival and necrosis. This 'apoptotic window' comprises a dose range where the damage is too serious to allow repair of the cell, but not severe enough to endanger the cellular ion homeostasis, a process, which usually leads to water influx and a loss of membrane integrity, both being typical of necrosis (see Fig. (2)) [37, 38]. Secondly, apoptosis has been described to have a rather anti-inflammatory effect, which is advantageous for cancer patients compared to massive necrosis and subsequent severe inflammation. However, the effects of apoptosis on the immune systems are being discussed at present, and the paradigm 'necrosis causes inflammation, apoptosis is anti-inflammatory' may be modified in close future [39-43].

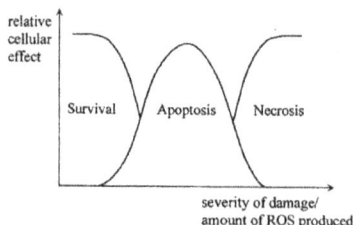

Fig. (2). Dependence of the cellular response on the increasing degree of PDT-induced damage. The basic modes of cellular response are dose-dependent and with increasing amount of ROS produced a transition from survival to apoptosis and finally to a necrotic response is usually observed.

While the mechanisms of apoptosis induction vary and are triggered via internal and external factors, the execution of apoptosis represents a typical process, which is – at least in many cases – not influenced by the initiation process itself

(for a detailed review on apoptosis see e.g. [44, 45]). Apoptosis is morphologically characterized by cell shrinkage and other distinctive changes such as nuclear chromatin condensation and segregation of the cell into apoptotic bodies, the latter preventing leaking of cell content (enzymes etc.) into tissue [46]. Among the characteristics of active cells death named above, other features involve changes in cellular biochemistry (e.g. the activation of a distinct class of proteases) and membrane structure; these hallmarks shall be discussed in the following sections.

Mitochondria: Central Organelles in Active Cell Death

Research during the last decade revealed a great range of apoptotic stimuli from inside or outside the cell to converge in mitochondria for further signal processing and execution. The intermembrane space of these organelles contains several proteins, which are centrally involved in apoptosis, among them cytochrome c, the apoptosis inducing factor (AIF), Smac/Diablo (second mitochondrial activator of caspases), Omi/Htr2A (high temperature requirement A) and inactive pro-forms of caspases [43]. Pores in the inner and/or outer mitochondrial membrane (IMM and OMM) execute the release of these proteins and fire off the apoptotic process. The pore-opening is further controlled by proteins of the Bcl-2 family (B-cell lymphoma 2, see below) [47-51].

Caspases: Inducers and Executers of Active Cell Death

Cysteinyl aspartic acid specific proteases (caspases) are hydrolytic enzymes centrally involved in the apoptotic process. The whole set of caspases is present in any cell in so-called pro-forms (zymogens, procaspases). Cleavage of the latter by autocatalysis or by other caspases activates their catalytic functions. Based on the fact that caspases may activate each other by proteolysis, it stands to reason that a cascade of caspases might be engaged during apoptosis (see also Fig. (3)). The most common activation pathway involves the processing of caspase 9 from procaspase 9 by the apoptosome, an enzyme complex which is compiled from cytochrome c, APAF-1 (apoptosis protease activation factor-

1) and (d)ATP ((deoxy-) adenosine-5´-triphosphate). Caspase 9 activates caspases 3, 6 and 7 ('effector' caspases) which attack several key enzymes on the metabolic map and manifest the morphological, biochemical and energetic changes typical of active cell death (a comprehensive list of physiological caspase substrates can be found in [52]). Caspases 2 and 8 (among others referred to as 'initiator' caspases) are involved in signal transmission for receptor-induced apoptosis (e.g. via the Fas- (CD95) or TNF receptor (tumor necrosis factor)) and can directly activate caspase 3. However, a cross-talk exists to pro-apoptotic members of the Bcl-2 family, which renders caspase 8 mediated apoptosis sensitive to mitochondrial control: caspase 8 cleaves Bid (a proapoptotic member of the Bcl-2 family), yielding tBid which translocates to mitochondria where it is assumed to be involved in mitochondrial pore opening and apoptosis execution [53-57].

The Conflictive Bcl-2 Protein Family

The proteins of the Bcl-2 family can regulate the onset of active cell death in various ways. Most of them act at the level of mitochondria by either advantaging or hampering pore formation (or opening). Bad, Bak and Bid represent pro-apoptotic proteins and are involved in the signal transmission from caspase 8 to mitochondria. At least Bax is known to be a component in some models of mitochondrial pores [58-60].

The anti-apoptotic proteins Bcl-2 and Bcl-X$_L$ block pore-opening and cytochrome c release. Other hypotheses suggest formation of pores by Bcl-2 and Bcl-X$_L$ and thus a proapoptotic function [61]. The regulation of pore opening remains unclear, but the ratio of pro- to anti-apoptotic Bcl-2 family members might be involved in the decision between survival and apoptosis [61-64]. For a recent overview on functions of Bcl-2 proteins, see [65].

Cell Surface Death Receptors

Receptors on the outer surface of the plasma membrane play an important role in the initiation of active cell death, especially in the effector functions of cellular immunity [66-68]. Two well described receptors are the Fas receptor (also known as Apo-1 or CD95) and the TNF receptor family. Binding of ligands to the respective receptor activates caspase 8 through the death inducing signalling complex (DISC, comprised of several adaptor molecules such as FADD (Fas associated death domain)) and may facultative involve mitochondrial control (the Bid crosstalk, see above and Fig. (3)). Several (parallel?) mechanisms of signal processing and transduction seem to exist in Fas- and TNF-triggered apoptosis that have been reviewed in [69-71] and [72-74], respectively.

Apoptosis and Cellular Energetics

The execution of apoptosis is an active process *per definitionem* and thus requires energy in form of equivalents of phosphorylation power. Key steps in the apoptotic cascade directly depend on the availability of (d)ATP or GTP (guanosine-5´-triphosphate). So, for example, the formation of the apoptosome from APAF-1 and cytochrome c [75, 76], the transport of pro-apoptotic factors into the nucleus [77], the condensation of chromatin and the formation of apoptotic

bodies [78] as well as morphological changes [79] require large amounts of nucleoside triphosphates. The decisive role of the intracellular ATP level was elucidated in several publications, where a minimum ATP level necessary for apoptosis execution was defined [80-82]. In accordance with these we have shown the intracellular ATP level to remain close to control levels until late in apoptosis, when active cell death was triggered by aluminium (III) phthalocyanine tetrasulfonate-based PDT (AlPcS$_4$-PDT) in A431 cells (human epidermoid carcinoma) [36]. Data from a subsequent publication of our group suggested glycolytic ATP formation to compensate for mitochondrial malfunction: the mitochondrial membrane potential – the most important driving force for ATP production – decreases early in the apoptotic process probably due to pore-opening in mitochondria. Cells grown in media without glucose, but with mitochondrial substrates (pyruvate for citric cycle / oxidative phosphorylation) failed to undergo apoptosis due to the lack of ATP; the cell death mode shifted towards necrosis [83]. This finding is substantiated by a recent publication by Kirveliene *et al.* [84]. Taken together, these results are indicative of the fact that apoptosis, as an active cellular process, not only requires minimal cellular energy levels but furthermore is character-ized by active mechanisms that ensure sufficient energy supply.

2. PARAMETERS INFLUENCING THE CELL DEATH MODE AFTER PDT

As PDT is known to effectively induce apoptosis, the occurrence of this mode of cell death, however, depends crucially on the severity of damage set in the cell and on the primary target of the photosensitizing agent. Apoptosis represents, as remarked above, a 'window' phenomenon: for any photosensitizer / cell type combination, a certain range of treatment parameters might exist, where cells die mainly by active cell death (according to Fig. (2)). This 'apoptotic window' may differ in its dimension, but it is to be found at treatment conditions in between survival and necrosis of the target cells, the latter causing an irreversible damage of cells [85]. The 'apoptotic window' for a given model system depends on several parameters. Among these, the treatment protocol, namely the incubation period and the concentration of the photosensitizing agent as well as the irradiation parameters represent factors, which can be influenced by the experimenter. Both account for the severity of damage set in the target cells. The incubation protocol determines the effective intracellular concentration of the photosensitizing agent; longer incubation periods and / or higher sensitizer concentrations result, in most cases, in higher intracellular levels up to a saturation value [85-87]. The dynamics of ROS produced (and therefore the damage) is further influenced by the irradiation parameters: light fluence [light energy per area] and light power density [light power per area]. Some authors report the oxygen available for ROS formation to be a limiting factor for ROS production at high power densities; accordingly, this limits the damage as well [88-90].

Not only the effective concentration and the irradiation influence the cell death mode, but also the chemical properties of the photosensitizing agent, since it determines both, the quantum yield and therefore the amount of ROS produced per mole photosensitizer and photons, but also the

1154   *Current Pharmaceutical Design, 2005, Vol. 11, No. 9*   *Plaetzer et al.*

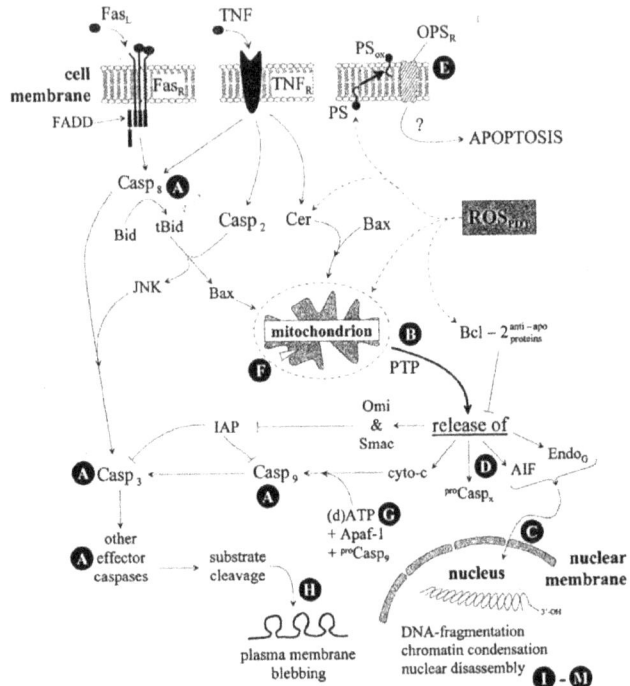

Fig. (3). Apoptotic pathways and -detection methods.

In the frame of the models of apoptosis induction (by PDT) the following methods can be applied for characterization and / or quantification of apoptotic cell death: the activity of several caspases is independently assessed by the use of synthetic fluorigenic (or colorimetric) caspase substrates (A, *fluorescence measurement* of reactions based on cell lysates or *flow cytometry*); mitochondrial membrane permeabilization may be visualized by calcein-AM stainings (B, *fluorescence microscopy*); translocation of EndoG and AIF can be demonstrated by immunochemical staining (C, visualization of protein localization by specific (fluorescent) antibodies – *fluorescence microscopy*); the latter technique is also applicable with proteins released from mitochondria (D; alternatively, subcellular fractions may be probed with antibodies by *western blotting*); exposure of (oxidized) phosphatidyl serine (PS) is determined by staining with fluorescence-labelled Annexin-V antibodies (E, *fluorescence microscopy* or *flow cytometry*); energetic changes during apoptotic cell death such as breakdown of the mitochondrial membrane potential (Δψ) and the intracellular ATP level are characterized by fluorescent stains (e.g. JC-1) and luciferase-based luminometric methods, respectively (F and G / *fluorescence microscopy* or *flow cytometry* and *luminometry*, respectively); formation of characteristic blebs (apoptotic bodies) of the cell membrane can be verified by *light* or *electron microscopy* (H).

The following apoptotic changes in the nucleus are accessible experimentally i) chromatin condensation and nuclear fragmentation are visualized by DNA-staining fluorescent dyes (I, *fluorescence microscopy*), ii) reduction of nuclear DNA content during fragmentation is shown by flow cytometry analysis (J, subG$_1$-assay), iii) DNA fragmentation can be further demonstrated by gel electrophoresis (K) or the COMET assay (L), and iv) the occurrence of free 3'-OH groups (a consequence of DNA fragmentation) can be detected using the TUNEL assay (M). For methodological details and appropriate references see text.

Apoptosis Following Photodynamic Tumor Therapy

Current Pharmaceutical Design, 2005, Vol. 11, No. 9 1155

cellular localization and thus the primary target of PDT [18, 19, 91]. The threshold concentration for complete cell killing therefore depends on the chemical properties as well. So, for example, we could find 100% cell kill of A431 cells by apoptosis / necrosis with 10 μM AlPcS₄ (a hydrophilic sensitizer) and fluences >4 J.cm^{-2} (660 nm). At similar fluences (at 610 nm), comparable cell kill could be found at concentrations of 0.09 μM hypericin (a lipophilic sensitizer from extracts of *hypericum perforatum*), about three orders of magnitude lower concentration than AlPcS₄ (unpublished results). In order to increase the efficiency of PDT, a wide spectrum of photosensitizing agents with different chemical properties has been successfully established during the last 25 years. They range from mainly lipophilic (such as hypericin, Photofrin and meso-tetrahydroxy phenylchlorine (mTHPC)) to hydrophilic (for example AlPcS₄). Due to their chemical structure, lipophilic dyes preferably localize in membrane systems, such as the plasma membrane, the ER (endoplasmatic reticulum) and Golgi systems, but also in mitochondria. On the contrary, hydrophilic dyes accumulate in the cytoplasm and lysosomes [16-19, 91]. Among all photosensitizers, δ-aminolevulinic acid- (ALA) induced endogenous protoporphyrin IX (PPIX, precursor in biosynthesis of heme) should be mentioned: an excess of ALA abolishes the negative feedback regulation of heme on its biosynthesis pathway and produces high concentrations of PPIX that can be employed as an endogenous photosensitizing agent [92-94]. The synthesis of PPIX runs in mitochondria and, as a consequence, ALA-derived PPIX will be selectively found in mitochondria, at least after short incubation periods [95]. The intimate connection between sensitizer localization and location of the primary damage is based on the short lifetime of ROS, which are responsible for the short diffusion lengths of these reactive molecules. Accordingly, Moan *et al.* have estimated the intracellular diffusion distance to be smaller than 20 nm [96]. Thus, the chemical properties clearly determine the area of primary damage: dyes localizing in the plasma membrane will therefore more likely cause a loss in membrane integrity (and subsequent necrosis) than dyes localizing in the cytoplasm, in mitochondria or other organelles [30, 97-100]. In the special case of ALA-induced PPIX, ROS formed in mitochondria have been shown to effectively induce apoptosis [101-104]. Consequently, in some cases, ROS formation by PDT in or in proximity of mitochondria has been shown to induce apoptosis in cells, which are otherwise resistant to apoptosis induction by conventional tumor treatment [34].

3. MECHANISMS OF PDT-INDUCED APOPTOSIS

The phenomenon of apoptosis induction by PDT has been depicted in an excellent review by Oleinick *et al.* [33]. Some possible mechanisms of how damage set by PDT triggers apoptosis (illustrated in Fig. (4)) shall be discussed in the following chapters.

3.1. Mitochondria as Primary PDT-Targets

Overproduction of ROS in Mitochondria and Its Direct Effects on Mitochondrial Pore Opening

Following a theory of Skulachev, which has been suggested in three outstanding review articles [3, 24, 105],

apoptosis represents an evolutionary old mechanism to eliminate ROS-overproducing cells, which pose a threat to the cellular community (tissue). Indeed, some of the key proteins (cytochrome c in the cytoplasm or the AIF) seem to play an important role in the cellular antioxidant defence. The primary role of these proteins might be to quench ROS in the cytoplasm. If the radicals exceed a certain level of quantity, the same proteins in the cytoplasm fire off active cell death [3, 24, 105].

PDT-produced ROS are likely to cause the same effect. For detailed mechanisms interrelating ROS production and the mitochondrial pore opening, several models have been suggested up to now (see Fig. (4)): oxidation of a critical cystein residue (cys^{56}) under conversion of sulfhydryl groups into disulfide bridges has been shown to transform the specific adenine nucleotide transporter (ANT, which imports ADP and exports ATP to the cytosol) into an unspecific pore permeable for solutes with less than 1.5 kDa [106, 107]. As a result, gradients of low molecular weight molecules between the matrix and the cytosol disappear. Normally, mainly concentration gradients of K$^+$ and Cl$^-$ regulate the osmotic pressure in the matrix, which are lost during pore opening. Now the matrix protein concentration, which is supposed to be higher than in the cytosol, is responsible for matrix osmolarity which, in turn, causes water influx and matrix swelling. The inner mitochondrial membrane is folded (cristae structure) and unfolds as a consequence of volume increase. Since the area of the outer membrane is smaller, its rupture leads to the release of pro-apoptotic proteins (e.g. cytochrome c and the AIF) from the intermembrane space into the cytoplasm which start the apoptotic cascade [108-111]. The effect of ROS-induced pore formation involving the ANT could be inhibited by dithiotreitol (DTT), a disulfide reductant [112]. Peroxidation of lipids surrounding the ANT is suggested to bring about the same effect [113, 114]. Damage of lipids in the mitochondrial membranes might cause an initial drop of the mitochondrial membrane potential (Δψ). This primary reduction in Δψ increases the probability of opening of the voltage dependent anion channel (VDAC) in the outer mitochondrial membrane due to the voltage-sensitivity of this channel [115-117]. Pore-opening via the VDAC allows escape of pro-apoptotic proteins and initiates the apoptotic cascade.

The essential participation of mitochondrial pores in PDT-induced apoptosis is also substantiated by findings which indicate that inhibitors of pore opening and -formation, such as cyclosporine A (CsA) and bongkrekic acid (BA) can cancel out apoptosis triggered by compounds that accumulate in mitochondria. So, for example, verteporfin has been shown to target the ANT and set off pore formation and Δψ loss in isolated mitochondria. This effect could be inhibited by CsA [118]. Singlet oxygen generated in mitochondria by Pc 4 PDT (phthalocyanine 4) on A431 cells caused a rapid permeabilization of the inner mitochondrial membrane, mitochondrial depolarization and cytochrome c release, as shown in a study by Lam *et al.* [117]. However, in clear contrast to these studies, irradiation of mitochondria loaded with hematoporphyrin rather prevented mitochondrial pore opening by a mechanism of site-selective inactivation of discrete pore functional domains [119].

1156   *Current Pharmaceutical Design, 2005, Vol. 11, No. 9*   *Plaetzer et al.*

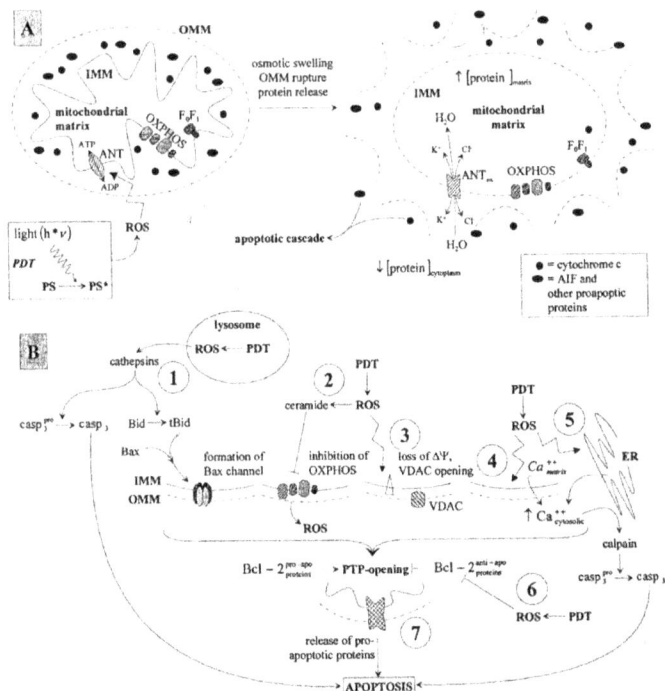

Fig. (4). Possible mechanisms of apoptosis induction by PDT.

A) PDT-generated ROS cause oxidation of a critical cystein residue the adenine nucleotide translocator (ANT) which thereby is transformed into an unspecific pore permeant for solutes less than 1.5 kDa (e.g. equilibration of the membrane gradient of chlorine and potassium ions). This causes osmotic water uptake due to the higher concentration of high molecular mass compounds in the mitochondrial matrix. As a result, the inner mitochondrial membrane (IMM) stretches causing rupture of the outer membrane (OMM). Pro-apoptotic proteins (cytochrome c and the apoptosis inducing factor (AIF) for example) are being released into the cytoplasm and initiate the apoptotic cascade. Oxidation of lipids surrounding the ANT might exert the same effect.

B) Detailed models of mitochondrial pore opening via PDT-produced ROS involve the release of cathepsins from lysosomes; the latter can either directly activate caspase 3 or cleave Bid (tBid) and induce mitochondrial pore opening via Bax (pathway indicated by 1). In the proximity of mitochondria, ceramide can be formed by ROS (pathway 2). This leads to inhibition of the oxidative phosphorylation, single electron transfer to oxygen and further (amplificated) formation of superoxide anions. The latter are potent inducers of mitochondrial pore opening. A decrease of the mitochondrial membrane potential due an ROS-induced loss of membrane integrity of the inner membrane leads to opening of the VDAC due to the voltage-sensitivity of this channel (pathway 3). An increase of mitochondrial Ca²⁺ as a consequence of ROS formation causes membrane destabilization, pore opening and apoptosis (pathway 4). Calcium released from the ER following PDT can either directly diffuse into mitochondria and induce pore opening via pathway 4 or directly activate caspase 3 through the calpain pathway (pathway 5). The anti-apoptotic protein Bcl-2 has been shown to be a target of PDT-formed ROS. Oxidation of Bcl-2 leads to pore opening (pathway 6). All these pathways converge into opening of more or less well defined mitochondrial pores which, in turn, allow escape of pro-apoptotic proteins (7). For references see text.

Apoptosis Following Photodynamic Tumor Therapy   *Current Pharmaceutical Design*, 2005, *Vol. 11, No. 9* 1157

Effects on Ceramide as Inducers of Mitochondrial Apoptosis

Ceramide (a sphingolipid) is known to play an important role in many initiation pathways of active cell death [120-122]. Separovic *et al.* reported PDT with Pc 4 to result in elevated ceramide levels in L5178Y-R cells and subsequent onset of active cell death [123, 124]. These findings were substantiated by Nagy *et al.* in A431 cells using the same photosensitizing agent [125, 126]. Wispriyono *et al.* also provided biochemical and genetic evidence of the role of the de novo synthesis of sphingolipids in apoptosis post Pc 4 PDT [127]. Pc 4 photodynamic therapy could even induce ceramide generation and apoptosis in acid sphingomyelinase-deficient mouse embryonic fibroblasts [128]. Production of ceramide could therefore represent a widespread mechanism of apoptosis induction by PDT.

Effects of Elevated Calcium Levels on Mitochondria

Calcium is a second messenger ion associated with a huge functional variety of regulatory processes in cells. Elevated calcium levels also accompany PDT-triggered apoptosis in many model systems (see below). It has been shown that mitochondrial Ca^{2+} overload leads to pore opening and onset of apoptosis [129-132]. Kowaltowsky *et al.* suggested Ca^{2+} ions to amplify the membrane depolarization induced by ROS by binding and destabilization of lipids on the inner surface of the mitochondrial membrane, thus leading to conformational changes in membrane proteins and pore opening [110, 133-135].

Effects on Anti-Apoptotic Proteins in Mitochondria

The anti-apoptotic protein Bcl-2 has been supposed to be a primary target of PDT, especially when the lipophilic photosensitizer Pc 4 is being used [136-138]. Overexpression of Bcl-2 protects from PDT-mediated apoptosis in several model systems and PDT with photosensitizers targeting Bcl-2, such as Pc 4 (which has a prevailing mitochondrial localization) can induce active cell death even in cells overexpressing Bcl-2; for details, see [33].

A recent publication reports that only membrane-inserted Bcl-2 is being photodamaged by light-excited Pc 4 since a cytosolic Bcl-2 mutant lacking the C-terminal domain for membrane insertion was not photodamaged [137, 138]. Photodamage set in the α-helices 5/6 of the region encompassing the BH1 and BH2 domains led to a crosslinking of the proteins. Photoactivated Pc 4 specifically destroyed Bcl-2, while other mitochondrial membrane proteins (such as Bak or the VDAC) or cytosolic proteins (such as Bax) were not affected [137, 138]. Treatment of HeLa cells with sublethal doses of hypericin PDT causes a phosphorylation of Bcl-2 at the Ser-70 residue [139]. The phosphorylation increased the cytoprotective role of Bcl-2, an effect, which has been proposed for other cell model systems as well [140]. However, the role of Bcl-2 seems to be dependent on the model system (photosensitizing agent and cell line used) and is hardly predictable at present.

3.2. Endoplasmatic Reticula as Primary PDT Targets

Often, PDT is accompanied by a significant increase in intracellular calcium. The latter may be released from internal stores (such as mitochondria or the ER) or come from outside the cell by influx through the plasma membrane [141]. Several authors correlated this increase in cytoplasmatic calcium with the onset of apoptosis. Tajiri *et al.* measured an strong increase of [Ca], 1-2 hours post Photofrin-PDT applied to HSC-2 squamous cell carcinomas, which was followed by apoptosis [142]. The coincidence of increased [Ca], and apoptosis was substantiated with PDT on V79 hamster cells [143]. Rueck *et al.* clearly identified a rise in [Ca], as a signal for apoptosis in rat bladder RR1002 cells treated with photoactivated AlPcS₄ [144]. For ALA-induced PPIX-PDT on HL 60 leukemia cells, Ca^{2+} release from the ER has been assumed to represent an alternative to mitochondrial induced apoptosis [103]. In this model, elevated levels of cytoplasmatic calcium were suggested to result from photodamage set to Ca^{2+}-binding proteins in the ER. The increase in [Ca], may contribute to apoptosis by activating the Ca^{2+}-calpain pathway, which cause PTP opening at mitochondria [133, 134] or induces activation of caspase 3 without mitochondrial involvement [145-147]. Additionally, fusion of the ER and mitochondria caused by ALA-PDT allows direct flow of calcium to mitochondria and leads to pore opening [103]. Bcl-2 has been reported to increase the emptying of ER calcium stores during apoptosis of Hela cells after verteporfin-PDT [148]. Thus, apoptosis induction after PDT using photosensitizers localizing in the ER is tightly connected with an increase of [Ca],, which either triggers mitochondrial pore opening or directly activates caspases.

3.3. Lysosomes as Primary PDT-Targets

Hydrophilic photosensitizers preferably localize in the cytoplasm and in lysosomes. Kessel *et al.* could show that lysosomal localizing photosensitizers cause the release of cathepsins from these organelles into the cytoplasm [149]. Once in the cytosol, cathepsins can either directly activate caspases [150, 151] or cleave Bid (yielding tBid) and promote this pathway leading to mitochondrial pore opening and apoptosis [152].

3.4. The Plasma Membrane as Primary PDT Target

Even photosensitizers localizing in the plasma membrane can trigger active cell death. Using evanescence illumination – which ensures photoactivation of sensitizing molecules only in the plasma membrane and a very small part of the cytoplasm – in combination with the photosensitizing drug Rose-Bengal, which localizes mainly in the plasma membrane, Lin *et al.* could observe apoptosis induction [153]; therefore, PDT-induced singlet oxygen production in the plasma membrane also can induce apoptosis. In this context, examination of the possible contribution of the outside phosphatidyl serine receptor (OPSR) to apoptotic signalling induced by ROS generated in the plasma membrane seems to be promising. The OPSR recognizes oxidized phosphatidyl serine on the outer leaflet of the plasma membrane and sends apoptotic signals to the cytoplasm [154, 155]. The contribution of the OPSR remains crucially hypothetical, but could prove worth explorating.

3.5. Other Mechanisms Leading to PDT-Induced Apoptosis

Several publications deal with the induction of expression and secretion of Fas and TNF-α by PDT. For

1158   *Current Pharmaceutical Design*, 2005, *Vol. 11, No. 9*   *Plaetzer et al.*

example, an increase of Fas in A431 cells and its appearance in the medium within one hour after PDT was measured by Ahmad *et al.* [156]. The data of Yslas *et al.* who also measured an increase of Fas antigen followed by apoptosis in PDT treated cells further substantiate the contribution of Fas to active cell death induced by PDT [157]. Pc 4 PDT applied to A431 leads to production of TNF-α, but its role in PDT-triggered apoptosis and the detailed mechanisms remain a topic of scientific discussion [158].

Also modifications of mitogen activated protein kinase (MAPK) cascades may represent a mechanism, by which PDT triggers induction of apoptosis. MAPKs are central mediators of the pathways regulating survival and cell death and have been implicated in the response of tumor cells to PDT [33]. While exposure of different cancer cell lines to PDT generally results in the activation of protein kinases JNKs and p38, which signal cellular survival, PDT causes inhibition of ERK kinases [159, 160]. The rapid down-regulation of the ERK pathway could be a necessary prerequisite for the induction of apoptosis after PDT. For details see [33].

4. APOPTOSIS DETECTION METHODS: BENEFITS AND DRAWBACKS

Several analytical methods utilize the differences in morphological, biochemical and energetic properties of (surviving,) necrotic and apoptotic cells to assess the modes of cell death. Some of these shall be described here, providing the principle of each assay and a brief evaluation of the expressiveness. In the context of the various mechanisms of apoptosis induction, these methods are outlined in Fig. (3). Since many assays make use of fluorescence dyes, the experimenter is advised to avoid misinterpretation of experimental data due to possible spectral interference with the photosensitizer. Since most of the mechanistic details of PDT-induced cell death have been described using *in vitro* systems we shall focus on the related methods. For a critical discussion of the methodology used with (fixed) tissue sections see [161].

4.1. Fluorescent Staining of the Nucleus

The late stages of apoptosis are characterized by several distinct changes in the cell nucleus; methods which assess these changes shall be discussed in the following chapter.

DAPI- and Hoechst-Staining for Nuclear Fragmentation

The dye 4', 6-diamidino-2-phenylindole, dihydrochloride (DAPI) as well as the bisbenzimide dyes Hoechst 33258, Hoechst 33342 and Hoechst 34580 represent classical nuclear and chromosome counterstains. They can be used to stain (fragmented) nuclei and detect apoptosis via fluorescence microscopy [32, 35, 162]. This approach provides single-cell analysis which additionally allows morphological analysis of the cells by microscopy. The major disadvantage is that a large number of cells have to be analysed in order to give a representative and significant quantification and the identification of fragmented nuclei is up to the investigator and therefore linked with subjective judgement.

Flow Cytometric Determination of the DNA Content / Cell Cycle Analysis

The method is carried out by measuring the DNA content of cells after fixation and permeabilization with ethanol and subsequent staining with propidium iodide (PI) using FACS analysis (fluorescence activated cell sorting) / flow cytometry. Cells in the resting and gap-phase (G_0/G_1) contain a certain amount of DNA (diploid phase), which is characterized by a distinct peak of red fluorescence. Due to nuclear fragmentation and condensation, the DNA content decreases during apoptosis, which can be quantified by the detection of cells with lower fluorescence, the so called subG$_1$ fraction [163-165]. Despite the general advantages of analysis by flow cytometry (single-cell analysis, rapid assessment of large cell numbers) the quantification of apoptosis at late stages of active cell death may lead to an underestimation since secondary necrosis may occur at that time (characterized by small cell fragments that are not recorded by this method). On the other hand, an overestimation of apoptotic cell death may occur due to counting of DNA-containing particles coming from one single cell as separate events.

DNA Laddering

Agarose gel electrophoresis is a method to detect fragmentation of DNA during apoptosis. DNA extracts of apoptotic cells (yielded by standard techniques of isolation of genomic DNA involving cell lysis and protein digestion by proteinase K, phenol-chloroform extraction and DNA precipitation by ethanol) show a typical 'DNA ladder' configuration consisting of multiples of 180 – 200 bp DNA-fragments when separated by agarose gel electrophoresis and visualized by ethidium bromide staining. These fragments are generated during late stages of apoptosis by Ca^{2+}/Mg^{2+}-dependent endonucleases which cleave the linker regions between nucleosomes [166, 167]. It should be mentioned that the detection of DNA fragmentation by gel electrophoresis is only qualitative and therefore *not* the method of choice for quantification of apoptosis in cell populations.

COMET-Assay

The single cell gel electrophoresis assay (also known as COMET assay) is a technique for analysing and measuring DNA damage/degradation in individual cells. Due to DNA fragmentation during apoptosis, this method is also applicable for detection of single apoptotic cells. In the COMET assay, cells are embedded in a thin agarose gel on a microscope slide. The cells are lysed to remove all cellular proteins and the DNA is subsequently allowed to unwind under alkaline or neutral conditions (depending on the protocol). In the following, the DNA is separated by electrophoresis, stained with a fluorescent dye (e.g. ethidium bromide) and cells are evaluated by fluorescence microscopy. During electrophoresis, DNA fragments migrate away from the nucleus resulting in a typical comet-like tail of apoptotic nuclei, while the DNA of intact cells remains in the nucleus [168]. Though allowing single-cell analysis, interpretation of the results might be difficult with respect to distinguishing between apoptosis and necrosis since (uncontrolled) DNA breaks may also occur during necrotic cell death [169].

Apoptosis Following Photodynamic Tumor Therapy

Current Pharmaceutical Design, 2005, *Vol. 11, No. 9* 1159

TUNEL Assay

The TUNEL technique (terminal deoxynucleotidyl transferase-mediated dUTP nick end labelling) is an enzymatic method for detection of DNA fragments generated during apoptosis. Free 3'-OH termini of DNA fragments are labelled by addition of FITC-dUTP (fluorescein isothiocyanate deoxy-uracil-5'-triphosphate) which is catalysed by terminal deoxynucleotide transferase (TdT) in a template-independent manner. FITC fluorescence is evaluated by flow cytometry analysis or fluorescence microscopy. Advantages of the TUNEL assay are the ability to reveal early DNA breaks during apoptosis and the good distinction of labelled and unlabelled cells [170-173]. According to a recent publication, care must be taken to avoid false-positive results when applying the TUNEL assay to tissue sections [174].

4.2. Determination of Biochemical Features of Apoptosis

Apoptosis, as active process, is characterized by several changes in cellular biochemistry and morphology. Some of the events typical of active cell death can be determined by enzymatic analysis; other methods deal with the change in localization of proteins relevant in the apoptotic process.

Caspase Assays

The activity of caspases, key players in apoptosis, can be analysed by the determination of cleavage of the specific fluorogenic (tetra-) peptide substrates [175, 176]. Several of these substrates, each more or less specific for one distinct caspase, are available commercially. Especially determination of caspase 3 activity is generally used as a specific marker for cells which have entered the execution phase of apoptosis. Briefly, cells are harvested, the protein content is being determined by standard methods, followed by lysis of cells. After addition of the substrate, the release and the concomitant activation of fluorochromes can be measured [57, 177, 178]. Recent developments allow flow cytometric analysis of caspase activation, thus providing single cell evaluation [179]. The activation of caspases involves the cleavage of inactive precursors (pro-caspases) and hence can be detected by western blot analysis using specific antibodies [180].

Staining of Phosphatidyl Serine on the Outer Surface of the Cell Membrane

In viable cells, phosphatidyl serine (PS) is directed to the cytoplasmatic compartment. However, in apoptotic cells, PS appears on the outer leaflet of the plasma membrane, where it can be detected by fluorescent labelled Annexin-V conjugates. Annexin-V probes are being used for flow cytometry analysis, confocal- or epifluorescence microscopy and can be combined with nucleic acid stains, such as propidium iodide (red fluorescence, permeant to necrotic cells only), to accurately assess mixed populations of viable (no fluorescence), apoptotic (Annexin-V fluorescence without red nuclear staining) and necrotic cells (ideally red fluorescence from nucleic stain only; under certain circumstances, also necrotic cells can display Annexin-V stain, possibly by binding to PS at the inner membrane surface) [181-184]. The use of this method requires accurate control procedures in order to avoid misinterpretation of PI stained secondary necrotic cells.

Determination of the Translocation of Proteins from Mitochondria into the Cytoplasm

The translocation of cytochrome c from the mitochondrial intermembrane space into the cytoplasm can be monitored by several methods. One approach involves subcellular fractionation and probing of cytosolic (and mitochondrial) fractions for the presence of cytochrome c by means of western blotting and immunochemical detection. This allows sensitive detection of cytochrome c in the cytoplasm but requires careful control procedures to prevent false-positive results caused by mitochondria damaged during the fractionation procedure. Using intact cells, the release of cytochrome c alternatively can be monitored by immunochemical visualization and fluorescence microscopy. For this purpose the cells are permeabilized and probed with anti-cytochrome c antibodies which, in turn, are labelled with secondary FITC-labelled antibodies. The leakage of cytochrome c is indicated by a change in the fluorescence pattern from punctuate signals to more diffuse fluorescence distributed in the cytoplasmatic compartment (and can be verified by costaining with mitochondria-specific dyes such as MitoTracker®) [185]. This method is of special interest when the temporal dynamics of cytochrome c release shall be analysed. A similar methodology is applicable also to analysis of translocation of AIF to the nucleus.

Determination of Pore Opening by Migration of Calcein-AM

Mitochondrial pore opening can be furthermore estimated by a quick and elegant method published by Lemasters *et al.* [186]: cells are loaded with calcein-AM, which is cleaved to fluorescent calcein by esterases in the cytoplasm. Calcein cannot translocate into mitochondria and those organelles may be counterstained by dyes such as TMRM (tetramethylrhodamine, methyl ester) as long as they produce some $\Delta\psi$. Whenever pore opening occurs during apoptosis, green calcein fluorescence becomes uniform within the cells due to diffusion of the dye into mitochondria, accompanied by a loss in red TMRM-flourescence caused by a loss of $\Delta\psi$. This alteration in fluorescence characteristics is monitored by confocal microscopy and (again) is well suited for time-course studies.

4.3. Determination of Energetic Parameters

Since apoptotic and necrotic cells show fundamental difference with respect to their energy metabolism, both modes of cell death can be identified by determination of some central energetic parameters.

Dynamics of Mitochondrial Metabolic Activity During Apoptosis

As a parameter for mitochondrial (metabolic) activity, the ability of the cells to convert the tetrazolium dye MTT (3-[4, 5-dimethylthiazol-2-yl]-2, 5-diphenyltetrazolium bromide) to formazan (catalysed by mitochondrial succinate dehydrogenase) can be measured and compared to that of untreated control cells [187]. We found the temporal dynamics of the MTT activity to perfectly reflect the apoptotic fraction of cells, when cells were treated with AlPcS$_4$ PDT using different light doses: at three hours post treatment, necrotic cells had lost the ability to reduce the formazan salt while

1160   *Current Pharmaceutical Design*, 2005, *Vol. 11, No. 9*   *Ploetzer et al.*

apoptotic cells, still metabolically active at this point in time, showed significant MTT activity. Twenty-four hours post treatment these cells had terminated apoptosis, therefore the MTT signal corresponded to the surviving fraction of cells. Therefore, the difference curve between these MTT signals recorded 3 and 24 hours post treatment is indicative of the 'apoptotic window' [36]; however, it is important to note that this assay cannot *per se* identify apoptotic cell death but with respect to the temporal dynamics it allows the fast and cheap assessment of the different cellular responses after PDT since the resulting blue formazan salt can easily be measured by microplate readers. Similar expressive results indicating the dose-dependent occurrence of different cellular responses were gained with other cell lines and photosensitizing agents (unpublished data of our group).

Staining of the Mitochondrial Membrane Potential (JC-1)

Due to opening of mitochondrial pores, a reduction of the mitochondrial membrane potential is a common feature of apoptosis. The latest generation of mitochondrial membrane potential sensitive dyes, such as 5, 5', 6, 6'-tetrachloro-1, 1', 3, 3'-tetraethylbenzimidazolyl-carbocyanine iodide (JC-1), undergoes a $\Delta\psi$-dependent accumulation in the mitochondrial matrix, which, in consequence, causes a shift in its emission maximum from green to red. Analysis can be done with flow cytometry or fluorescence microscopy [188, 189]. Untreated control cells show high red fluorescence, whereas the green signal is low. Necrotic cells are characterized by a *drastic* reduction of the red fluorescence and high green fluorescence. Apoptotic cells are characterized by a *lower*, but not completely collapsed mitochondrial membrane potential [109, 190]. The red signal is about 5 times lower than that of the untreated controls, while the green signal appears to be slightly increased [83]. As discussed by Bernardi *et al.* [191], only the red fluorescence of JC-1 (and not the ratio of green-to-red) is indicative of mitochondrial $\Delta\psi$.

Time Course of the Intracellular ATP Level

The intracellular ATP level has been reported to be a significant determinant of the cell death mode [81, 82, 192]. Apoptotic cells maintain high ATP levels for several hours, whereas necrotic cells are characterized by a rapid and complete loss of intracellular ATP [36]. For measurement, ATP is liberated from cells by either rapid lysis or extraction of the nucleotides by a chaotropic agent (perchloric or trichloroacetic acid). Once released from the cells, it can be determined by a luminescence reaction based on the firefly luciferase-catalysed ATP-dependent oxidation of D-luciferin [193]. Two methods for the rapid determination of intracellular ATP we introduced in [194] with special respect to the use of the 96 well microplate format which allows rapid and easy handling of high sample numbers.

CONCLUSIONS

(i) Photodynamic therapy is efficient in triggering apoptosis in target cells. This might be due to the main effect of PDT, the formation of ROS, either directly in mitochondria or in their proximity.

(ii) There seems to be no universal mechanism of PDT. The pathway leading to apoptosis induction depends on the

model system and is tightly connected with the localization of the photosensitizing agent, which – in turn – determines the area of primary damage. As a common denominator, the involvement of mitochondria seems to be central to the apoptotic process induced by most variations of PDT.

(iii) PDT seems to be able to overcome the blocks to cell killing by induction of apoptosis in tumor cells resistant to therapies with ionizing radiation or chemotherapeutics. One might hypothesize that this is because of PDT triggers apoptosis at a 'late' stage within the apoptotic signal transduction pathway(s), in some cases by directly inducing mitochondrial pore-opening (or by photodamaging anti-apoptotic Bcl-2 proteins).

(iv) Numerous methods exist that allow the qualitative and – in some cases – also quantitative assessment of cell death. These methods are set apart from each other by their ability to accurately distinguish between the modes of cell death (apoptosis versus necrosis) and by their applicability to different kinds of samples. It is of fundamental importance for meaningful interpretation of experimental data and results published concerning the detection of cell death to be aware of the limitations and drawbacks each method is characterized by.

It can be summarized that the research on PDT-induced active cell death – as in part cited in the present review – has essentially contributed to the comprehension of the underlying cellular mechanisms. Although many of the molecular details remain unclear we are convinced that additional basic research unravelling regulation and the dynamics of the cellular response following PDT will further enhance the efficiency of photodynamic treatment.

ACKNOWLEDGEMENTS

We would like to thank Juergen Berlanda and Franz Obermair for fruitful discussion and for proof-reading the manuscript. We furthermore apologize to all authors, who contributed to the understanding of apoptosis following PDT and were not cited here.

ABBREVIATIONS

(d)ATP	=	(Deoxy) adenosine-5'-triphosphate
AIF	=	Apoptosis inducing factor
ALA	=	δ-Aminolevulinic acid
AlPcS$_4$	=	Aluminum (III) phthalocyanine tetrasulfonate
ANT	=	Adenine nucleotide translocator
APAF-1	=	Apoptosis protease activation factor-1
BA	=	Bongkrekic acid
Bcl-2	=	B-cell lymphoma 2
caspases	=	Cysteinyl aspartic acid specific proteases
COMET	=	Single cell gel electrophoresis assay
CsA	=	Cyclosporine A
$\Delta\psi$	=	Mitochondrial membrane potential
DAPI	=	4', 6-diamidino-2-phenylindole dihydrochloride

DIABLO	=	See Smac
DISC	=	Death inducing signalling complex
DTT	=	Dithiotreitol
ER	=	Endoplasmatic reticulum
FACS	=	Fluorescence activating cell sorting, flow cytometry
FADD	=	Fas associated death domain
FITC-dUTP	=	Fluorescein isothiocyanate deoxy-uracil-5'-triphosphate
GTP	=	Guanosine-5'-triphosphate
H_2O_2	=	Hydrogen peroxide
Htr2A/Omi	=	High temperature requirement A
IAP	=	Inhibitor of apoptosis proteins
IMM	=	Inner mitochondrial membrane
JC-1	=	5, 5', 6, 6'-tetrachloro-1, 1', 3, 3'-tetraethylbenzimidazolyl-carbocyanine iodide
MAPK	=	Mitogen activated protein kinase
mTHPC	=	Meso-tetrahydroxy phenylchlorine
MTT	=	3-[4, 5-dimethylthiazol-2-yl]-2, 5-diphenyltetrazolium bromide
NF-κB	=	Nuclear factor κ-B
Omi	=	See Htr2A
OMM	=	Outer mitochondrial membrane
OPSR	=	Outside phosphatidyl serine receptor
Pc 4	=	Phthalocyanine 4
PDT	=	Photodynamic therapy
PI	=	Propidium iodide
PPIX	=	Protoporphyrin IX
PS	=	Phosphatidyl serine
ROS	=	Reactive oxygen species
Smac/DIABLO	=	Second mitochondrial activator of caspases
tBid	=	Truncated Bid
TdT	=	Terminal deoxynucleotide transferase
TMRM	=	Tetramethylrhodamine, methyl ester
TNF	=	Tumor necrosis factor
TUNEL	=	Terminal deoxynucleotidyl transferase-mediated dUTP nick end labelling
VDAC	=	Voltage-dependent anion channel

REFERENCES

References 195-197 are related articles recently published in Current Pharmaceutical Design.

[1]   Raha S, Robinson BH. Mitochondria, oxygen free radicals, and apoptosis. Am J Med Genet 2001; 106 (1): 62-70.

[2]   Simon HU, Haj-Yehia A, Levi-Schaffer F. Role of reactive oxygen species (ROS) in apoptosis induction. Apoptosis 2000; 5 (5): 415-8.

[3]   Skulachev VP. Mitochondria in the programmed death phenomena; a principle of biology: "it is better to die than to be wrong". IUBMB Life 2000; 49 (5): 365-73.

[4]   Dougherty TJ. An update on photodynamic therapy applications. J Clin Laser Med Surg 2002; 20 (1): 3-7.

[5]   Hsi RA, Rosenthal DI, Glatstein E. Photodynamic therapy in the treatment of cancer: current state of the art. Drugs 1999; 57 (5): 725-34.

[6]   Chang SC, Bown SG. Photodynamic therapy: applications in bladder cancer and other malignancies. J Formos Med Assoc 1997; 96 (11): 853-63.

[7]   Levy JG, Obochi M. New applications in photodynamic therapy. Introduction. Photochem Photobiol 1996; 64 (5): 737-9.

[8]   Jori G. Photosensitized processes *in vivo*: proposed phototherapeutic applications. Photochem Photobiol 1990; 52 (2): 439-43.

[9]   Lin CW. Photodynamic therapy of malignant tumors--recent developments. Cancer Cells 1991; 3 (11): 437-44.

[10]   Hornung R, *et al.* Highly selective targeting of ovarian cancer with the photosensitizer PEG-m-THPC in a rat model. Photochem Photobiol 1999; 70 (4): 624-9.

[11]   Friberg EG, Cunderlikova B, Pettersen EO, Moan J. pH effects on the cellular uptake of four photosensitizing drugs evaluated for use in photodynamic therapy of cancer. Cancer Lett 2003; 195 (1): 73-80.

[12]   Bellnier DA, Young DN, Detty MR, Camacho SH, Oseroff AR. pH-dependent chalcogenopyrylium dyes as potential sensitizers for photodynamic therapy: selective retention in tumors by exploiting pH differences between tumor and normal tissue. Photochem Photobiol 1999; 70 (4): 630-6.

[13]   Henderson BW, Dougherty TJ. How does photodynamic therapy work? Photochem Photobiol 1992; 55 (1): 145-57.

[14]   Dougherty TJ, *et al.* Photodynamic therapy. J Natl Cancer Inst 1998; 90 (12): 889-905.

[15]   Dougherty TJ. Photodynamic therapy. Photochem Photobiol 1993; 58 (6): 895-900.

[16]   Boyle RW, Dolphin D. Structure and biodistribution relationships of photodynamic sensitizers. Photochem Photobiol 1996; 64 (3): 469-85.

[17]   Spikes JD. Chlorins as photosensitizers in biology and medicine. J Photochem Photobiol B 1990; 6 (3): 259-74.

[18]   Heier SK, Heier LM. Tissue sensitizers. Gastrointest Endosc Clin N Am 1994; 4 (2): 327-52.

[19]   Moan J. Properties for optimal PDT sensitizers. J Photochem Photobiol B 1990; 5 (3-4): 521-4.

[20]   Pushpan SK, *et al.* Porphyrins in photodynamic therapy - a search for ideal photosensitizers. Curr Med Chem Anti-Canc Agents 2002; 2 (2): 187-207.

[21]   Takemura T, Ohta N, Nakajima S, Sakata I. Critical importance of the triplet lifetime of photosensitizer in photodynamic therapy of tumor. Photochem Photobiol 1989; 50 (3): 339-44.

[22]   Halliwell B, Gutteridge JM. Free radicals, lipid peroxidation, and cell damage. Lancet 1984; 2 (8411): 1095.

[23]   Halliwell B, Gutteridge JM. Oxygen toxicity, oxygen radicals, transition metals and disease. Biochem J 1984; 219 (1): 1-14.

[24]   Skulachev VP. Mitochondrial physiology and pathology; concepts of programmed death of organelles, cells and organisms. Mol Aspects Med 1999; 20 (3): 139-84.

[25]   Agarwal R, Athar M, Bickers DR, Mukhtar H. Evidence for the involvement of singlet oxygen in the photodestruction by chloroaluminum phthalocyanine tetrasulfonate. Biochem Biophys Res Commun 1990; 173 (1): 34-41.

[26]   Godar DE. Light and death: photons and apoptosis. J Investig Dermatol Symp Proc 1999; 4 (1): 17-23.

[27]   Ochsner M. Photophysical and photobiological processes in the photodynamic therapy of tumours. J Photochem Photobiol B 1997; 39 (1): 1-18.

[28]   Weishaupt KR, Gomer CJ, Dougherty TJ. Identification of singlet oxygen as the cytotoxic agent in photoinactivation of a murine tumor. Cancer Res 1976; 36 (7 PT 1): 2326-9.

[29]   Plaetzer K, Kiesslich T, Verwanger T, Krammer B. The Modes of Cell Death Induced by PDT: An Overview. Medical Laser Application 2003; 18 (1): 7-19.

[30]   Kessel D, Luo Y. Photodynamic therapy: A mitochondrial inducer of apoptosis. Cell Death Differentiation 1999; 6 (1): 28-35.

1162   *Current Pharmaceutical Design, 2005, Vol. 11, No. 9*   *Plaetzer et al.*

[31] Moor AC. Signaling pathways in cell death and survival after photodynamic therapy. J Photochem Photobiol B 2000; 57 (1): 1-13.

[32] Luo Y, Chang CK, Kessel D. Rapid initiation of apoptosis by photodynamic therapy. Photochem Photobiol 1996. 63 (4). 528-34.

[33] Oleinick NL, Morris RL, Belichenko I. The role of apoptosis in response to photodynamic therapy: what, where, why, and how. Photochem Photobiol Sci 2002; 1 (1): 1-21.

[34] Stewart F, Baas P, Star W. What does photodynamic therapy have to offer radiation oncologists (or their cancer patients)? Radiother Oncol 1998; 48 (3): 233-48.

[35] Luo Y, Kessel D. Initiation of apoptosis versus necrosis by photodynamic therapy with chloroaluminum phthalocyanine. Photochem Photobiol 1997; 66 (4): 479-83.

[36] Plaetzer K, Kiesslich T, Krammer B, Hammerl P. Characterization of the cell death modes and the associated changes in cellular energy supply in response to AlPcS4-PDT. Photochem Photobiol Sci 2002; 1 (3): 172-177.

[37] Zhou CN. Mechanisms of tumor necrosis induced by photodynamic therapy. J Photochem Photobiol B 1989; 3 (3): 299-318.

[38] Buja LM, Eigenbrodt ML, Eigenbrodt EH. Apoptosis and necrosis. Basic types and mechanisms of cell death. Arch Pathol Lab Med 1993; 117 (12): 1208-14.

[39] Reiter I, Krammer B, Schwamberger G. Cutting edge: differential effect of apoptotic versus necrotic tumor cells on macrophage antitumor activities. J Immunol 1999; 163 (4): 1730-2.

[40] Cecic I, Parkins CS, Korbelik M. Induction of systemic neutrophil response in mice by photodynamic therapy of solid tumors. Photochem Photobiol 2001; 74 (5): 712-20.

[41] Korbelik M, Dougherty GJ. Photodynamic therapy-mediated immune response against subcutaneous mouse tumors. Cancer Res 1999; 59 (8): 1941-6.

[42] Korbelik M. Induction of tumor immunity by photodynamic therapy. J Clin Laser Med Surg 1996; 14 (5): 329-34.

[43] Sun J, Cecic I, Parkins CS, Korbelik M. Neutrophils as inflammatory and immune effectors in photodynamic therapy-treated mouse SCCVII tumours. Photochem Photobiol Sci 2002; 1 (9): 690-5.

[44] Zimmermann KC, Green DR. How cells die: apoptosis pathways. J Allergy Clin Immunol 2001; 108 (4 Suppl): S99-103.

[45] Zimmermann KC, Bonzon C, Green DR. The machinery of programmed cell death. Pharmacol Ther 2001; 92 (1): 57-70.

[46] Hacker G. The morphology of apoptosis. Cell Tissue Res 2000; 301 (1): 5-17.

[47] van Loo G, Saelens X, van Gurp M, MacFarlane M, Martin SJ, Vandenabeele P. The role of mitochondrial factors in apoptosis: a Russian roulette with more than one bullet. Cell Death Differ 2002; 9 (10): 1031-42.

[48] Gottlieb RA. Mitochondria: execution central. FEBS Lett 2000; 482 (1-2): 6-12.

[49] Green DR, Reed JC. Mitochondria and apoptosis. Science 1998; 281 (5381): 1309-12.

[50] Susin SA, Zamzami N, Kroemer G. Mitochondria as regulators of apoptosis: doubt no more. Biochim Biophys Acta 1998; 1366 (1-2): 151-65.

[51] Waterhouse NJ, Ricci JE, Green DR. And all of a sudden it's over: mitochondrial outer-membrane permeabilization in apoptosis. Biochimie 2002; 84 (2-3): 113-21.

[52] Fischer U, Janicke RU, Schulze-Osthoff K. Many cuts to ruin: a comprehensive update of caspase substrates. Cell Death Differ 2003; 10 (1): 76-100.

[53] Thornberry NA, Lazebnik Y. Caspases: enemies within. Science 1998; 281 (5381): 1312-6.

[54] Villa P, Kaufmann SH, Earnshaw WC. Caspases and caspase inhibitors. Trends Biochem Sci 1997; 22 (10): 388-93.

[55] Stennicke HR, Salvesen GS. Properties of the caspases. Biochim Biophys Acta 1998; 1387 (1-2): 17-31.

[56] Stegh AH, Peter ME. Apoptosis and caspases. Cardiol Clin 2001; 19 (1): 13-29.

[57] Earnshaw WC, Martins LM, Kaufmann SH. Mammalian caspases: structure, activation, substrates, and functions during apoptosis. Annu Rev Biochem 1999; 68: 383-424.

[58] Marzo I, et al. Bax and adenine nucleotide translocator cooperate in the mitochondrial control of apoptosis. Science 1998; 281 (5385): 2027-31.

[59] Brenner C, et al. Bcl-2 and Bax regulate the channel activity of the mitochondrial adenine nucleotide translocator. Oncogene 2000; 19 (3): 329-36.

[60] Belzacq AS, Vieira HL, Kroemer G, Brenner C. The adenine nucleotide translocator in apoptosis. Biochimie 2002; 84 (2-3): 167-76.

[61] Schendel SL, Montal M, Reed JC. Bcl-2 family proteins as ion-channels. Cell Death Differ 1998; 5 (5): 372-80.

[62] Antonsson B, Martinou JC. The Bcl-2 protein family. Exp Cell Res 2000; 256 (1): 50-7.

[63] Gross A, McDonnell JM, Korsmeyer SJ. BCL-2 family members and the mitochondria in apoptosis. Genes Dev 1999; 13 (15): 1899-911.

[64] Tsujimoto Y, Shimizu S. Bcl-2 family: life-or-death switch. FEBS Lett 2000; 466 (1): 6-10.

[65] Borner C. The Bcl-2 protein family: sensors and checkpoints for life-or-death decisions. Mol Immunol 2003; 39 (11): 615-47.

[66] Berke G. Killing mechanisms of cytotoxic lymphocytes. Curr Opin Hematol 1997; 4 (1): 32-40.

[67] Shresta S, Pham CT, Thomas DA, Graubert TA, Ley TJ. How do cytotoxic lymphocytes kill their targets? Curr Opin Immunol 1998; 10 (5): 581-7.

[68] Waring P, Mullbacher A. Cell death induced by the Fas/Fas ligand pathway and its role in pathology. Immunol Cell Biol 1999; 77 (4): 312-7.

[69] Chinnaiyan AM, Dixit VM. Portrait of an executioner: the molecular mechanism of FAS/APO-1-induced apoptosis. Semin Immunol 1997; 9 (1): 69-76.

[70] Peter ME, Krammer PH. Mechanisms of CD95 (APO-1/Fas)-mediated apoptosis. Curr Opin Immunol 1998; 10 (5): 545-51.

[71] Wajant H. The Fas signaling pathway: more than a paradigm. Science 2002; 296 (5573): 1635-6.

[72] Abe K, Kurakin A, Mohseni-Maybodi M, Kay B, Khosravi-Far R. The complexity of TNF-related apoptosis-inducing ligand. Ann N Y Acad Sci 2000; 926: 52-63.

[73] Rath PC, Aggarwal BB. TNF-induced signaling in apoptosis. J Clin Immunol 1999; 19 (6): 350-64.

[74] Wallach D, Boldin M, Varfolomeev E, Beyaert R, Vandenabeele P, Fiers W. Cell death induction by receptors of the TNF family: towards a molecular understanding. FEBS Lett 1997; 410 (1): 96-106.

[75] Liu X, Kim CN, Yang J, Jemmerson R, Wang X. Induction of apoptotic program in cell-free extracts: requirement for dATP and cytochrome c. Cell 1996; 86 (1): 147-57.

[76] Zou H, Li Y, Liu X, Wang X. An APAF-1.cytochrome c multimeric complex is a functional apoptosome that activates procaspase-9. J Biol Chem 1999; 274 (17): 11549-56.

[77] Yasuhara N, Eguchi Y, Tachibana T, Imamoto N, Yoneda Y, Tsujimoto Y. Essential role of active nuclear transport in apoptosis. Genes Cells 1997; 2 (1): 55-64.

[78] Kass GE, Eriksson JE, Weis M, Orrenius S, Chow SC. Chromatin condensation during apoptosis requires ATP. Biochem J 1996; 318 (Pt 3): 749-52.

[79] Mills JC, Stone NL, Pittman RN. Extranuclear apoptosis. The role of the cytoplasm in the execution phase. J Cell Biol 1999; 146 (4): 703-8.

[80] Nicotera P, Leist M, Ferrando-May E. Intracellular ATP, a switch in the decision between apoptosis and necrosis. Toxicol Lett 1998; 102-103: 139-42.

[81] Eguchi Y, Shimizu S, Tsujimoto Y. Intracellular ATP levels determine cell death fate by apoptosis or necrosis. Cancer Res 1997; 57 (10): 1835-40.

[82] Leist M, Single B, Castoldi AF, Kuhnle S, Nicotera P. Intracellular adenosine triphosphate (ATP) concentration: a switch in the decision between apoptosis and necrosis. J Exp Med 1997: 185 (8): 1481-6.

[83] Oberdanner CB, Kiesslich T, Krammer B, Plaetzer K. Glucose is required to maintain high ATP-levels for the energy-utilizing steps during PDT-induced apoptosis. Photochem Photobiol 2002, 76 (6): 695-703.

[84] Kirveliene V, Sadauskaite A, Kadziauskas J, Sasnauskiene S, Juodka B. Correlation of death modes of photosensitized cells with intracellular ATP concentration. FEBS Lett 2003; 553 (1-2): 167-72.

Apoptosis Following Photodynamic Tumor Therapy   *Current Pharmaceutical Design, 2005, Vol. 11, No. 9*   1163

[85]   Wyld L, Reed MW, Brown NJ. Differential cell death response to photodynamic therapy is dependent on dose and cell type. Br J Cancer 2001; 84 (10): 1384-6.

[86]   Ris HB, *et al.* Effect of drug-light interval on photodynamic therapy with meta-tetrahydroxyphenylchlorin in malignant mesothelioma. Int J Cancer 1993; 53 (1): 141-6.

[87]   Grubinger M, Hammerl P, Banieghbal E, Krammer B. Accumulation of aminolevulinic acid-induced protoporphyrin IX as photosensitizer in L-929 cells. Research Advances in Photochemistry and Photobiology 2000; 1: 137-145.

[88]   Henderson BW, Fingar VH. Relationship of tumor hypoxia and response to photodynamic treatment in an experimental mouse tumor. Cancer Res 1987; 47 (12): 3110-4.

[89]   Sitnik TM, Hampton JA, Henderson BW. Reduction of tumour oxygenation during and after photodynamic therapy in vivo: effects of fluence rate. Br J Cancer 1998; 77 (9): 1386-94.

[90]   Henderson BW, Fingar VH. Oxygen limitation of direct tumor cell kill during photodynamic treatment of a murine tumor model. Photochem Photobiol 1989; 49 (3): 299-304.

[91]   Moan J, *et al.* Intracellular localization of photosensitizers. Ciba Found Symp 1989; 146: 95-107; discussion 107-11.

[92]   Peng Q, *et al.* 5-Aminolevulinic acid-based photodynamic therapy. Clinical research and future challenges. Cancer 1997; 79 (12): 2282-308.

[93]   Taylor EL, Brown SB. The advantages of aminolevulinic acid photodynamic therapy in dermatology. J Dermatolog Treat 2002; 13 Suppl 1: S3-11.

[94]   Morton CA. The emerging role of 5-ALA-PDT in dermatology: is PDT superior to standard treatments? J Dermatolog Treat 2002; 13 Suppl 1: S25-9.

[95]   Uberriegler KP, Banieghbal E, Krammer B. Subcellular damage kinetics within co-cultivated WI38 and VA13-transformed WI38 human fibroblasts following 5-aminolevulinic acid-induced protoporphyrin IX formation. Photochem Photobiol 1995; 62 (6): 1052-7.

[96]   Moan J, Berg K. The photodegradation of porphyrins in cells can be used to estimate the lifetime of singlet oxygen. Photochem Photobiol 1991; 53 (4): 549-53.

[97]   Gomer CJ, Luna M, Ferrario A, Wong S, Fisher AM, Rucker N. Cellular targets and molecular responses associated with photodynamic therapy. J Clin Laser Med Surg 1996; 14 (5): 315-21.

[98]   Henderson BW, Bellnier DA. Tissue localization of photosensitizers and the mechanism of photodynamic tissue destruction. Ciba Found Symp 1989; 146: 112-25; discussion 125-30.

[99]   Oleinick NL, Evans HH. The photobiology of photodynamic therapy: cellular targets and mechanisms. Radiat Res 1998; 150 (5 Suppl) S146-56.

[100] Peng Q, Moan J, Nesland JM. Correlation of subcellular and intratumoral photosensitizer localization with ultrastructural features after photodynamic therapy. Ultrastruct Pathol 1996; 20 (2): 109-129.

[101] Gad F, Viau G, Boushira M, Bertrand R, Bissonnette R. Photodynamic therapy with 5-aminolevulinic acid induces apoptosis and caspase activation in malignant T cells. J Cutan Med Surg 2001; 5 (1): 8-13.

[102] Bourre L, Rousset N, Thibaut S, Eleouet S, Lajat Y, Patrice T. PDT effects of m-THPC and ALA, phototoxicity and apoptosis. Apoptosis 2002; 7 (3): 221-30.

[103] Grebenova D, *et al.* Mitochondrial and endoplasmic reticulum stress-induced apoptotic pathways are activated by 5-aminolevulinic acid-based photodynamic therapy in HL60 leukemia cells. J Photochem Photobiol B 2003; 69 (2): 71-85.

[104] Noodt BB, Berg K, Stokke T, Peng Q, Nesland JM. Apoptosis and necrosis induced with light and 5-aminolaevulinic acid-derived protoporphyrin IX. Br J Cancer 1996; 74 (1): 22-9.

[105] Skulachev VP. The programmed death phenomena, aging, and the Samurai law of biology. Exp Gerontol 2001; 36 (7): 995-1024.

[106] Costantini P, *et al.* Oxidation of a critical thiol residue of the adenine nucleotide translocator enforces Bcl-2-independent permeability transition pore opening and apoptosis. Oncogene 2000; 19 (2): 307-14.

[107] Moreno G, Poussin K, Ricchelli F, Salet C. The effects of singlet oxygen produced by photodynamic action on the mitochondrial

permeability transition differ in accordance with the localization of the sensitizer. Arch Biochem Biophys 2001; 386 (2): 243-50.

[108] Halestrap AP, McStay GP, Clarke SJ. The permeability transition pore complex: another view. Biochimie 2002; 84 (2-3): 153-66.

[109] Chiu SM, Oleinick NL. Dissociation of mitochondrial depolarization from cytochrome c release during apoptosis induced by photodynamic therapy. Br J Cancer 2001; 84 (8): 1099-106.

[110] Kowaltowski AJ, Castilho RF, Vercesi AE. Mitochondrial permeability transition and oxidative stress. FEBS Lett 2001; 495 (1-2): 12-5.

[111] Martinou JC, Green DR. Breaking the mitochondrial barrier. Nat Rev Mol Cell Biol 2001; 2 (1): 63-7.

[112] Costantini P, Colonna R, Bernardi P. Induction of the mitochondrial permeability transition by N-ethylmaleimide depends on secondary oxidation of critical thiol groups. Potentiation by copper-ortho-phenanthroline without dimerization of the adenine nucleotide translocase. Biochim Biophys Acta 1998; 1365 (3): 385-92.

[113] Augustin W, Gellerich F, Wiswedel I, Evtodienko Y, Zinchenko V. Inhibition of cation efflux by antioxidants during oscillatory ion transport in mitochondria. FEBS Lett 1979; 107 (1): 151-4.

[114] Lenaz G. Role of mitochondria in oxidative stress and ageing. Biochim Biophys Acta 1998; 1366 (1-2): 53-67.

[115] Bernardi P. Modulation of the mitochondrial cyclosporin A-sensitive permeability transition pore by the proton electrochemical gradient. Evidence that the pore can be opened by membrane depolarization. J Biol Chem 1992; 267 (13): 8834-9.

[116] Bernardi P. Mitochondrial transport of cations: channels, exchangers, and permeability transition. Physiol Rev 1999; 79 (4): 1127-55.

[117] Lam M, Oleinick NL, Nieminen AL. Photodynamic therapy-induced apoptosis in epidermoid carcinoma cells. Reactive oxygen species and mitochondrial inner membrane permeabilization. J Biol Chem 2001; 276 (50): 47379-86.

[118] Belzacq AS, *et al.* Apoptosis induction by the photosensitizer verteporfin: identification of mitochondrial adenine nucleotide translocator as a critical target. Cancer Res 2001; 61 (4): 1260-4.

[119] Salet C, Moreno G, Ricchelli F, Bernardi P. Singlet oxygen produced by photodynamic action causes inactivation of the mitochondrial permeability transition pore. J Biol Chem 1997; 272 (35): 21938-43.

[120] Birbes H, Bawab SE, Obeid LM, Hannun YA. Mitochondria and ceramide: intertwined roles in regulation of apoptosis. Adv Enzyme Regul 2002; 42: 113-29.

[121] Hannun YA, Obeid LM. Mechanisms of ceramide-mediated apoptosis. Adv Exp Med Biol 1997; 407: 145-9.

[122] Pettus BJ, Chalfant CE, Hannun YA. Ceramide in apoptosis: an overview and current perspectives. Biochim Biophys Acta 2002; 1585 (2-3): 114-25.

[123] Separovic D, Mann KJ, Oleinick NL. Association of ceramide accumulation with photodynamic treatment-induced cell death. Photochem Photobiol 1998; 68 (1): 101-9.

[124] Separovic D, He J, Oleinick NL. Ceramide generation in response to photodynamic treatment of L5178Y mouse lymphoma cells. Cancer Res 1997; 57 (9): 1717-21.

[125] Nagy B, Yeh WC, Mak TW, Chiu SM, Separovic D. FADD null mouse embryonic fibroblasts undergo apoptosis after photosensitization with the silicon phthalocyanine Pc 4. Arch Biochem Biophys 2001; 385 (1): 194-202.

[126] Nagy B, Chiu SM, Separovic D. Fumonisin B1 does not prevent apoptosis in A431 human epidermoid carcinoma cells after photosensitization with a silicon phthalocyanine. J Photochem Photobiol B 2000; 57 (2-3): 132-41.

[127] Wispriyono B, Schmelz E, Pelayo H, Hanada K, Separovic D. A role for the novo sphingolipids in apoptosis of photosensitized cells. Exp Cell Res 2002; 279 (1): 153-65.

[128] Chiu SM, Davis TW, Meyers M, Ahmad N, Mukhtar H, Separovic D. Phthalocyanine 4-photodynamic therapy induces ceramide generation and apoptosis in acid sphingomyelinase-deficient mouse embryonic fibroblasts. Int J Oncol 2000; 16 (2): 423-7.

[129] Chakraborti T, Das S, Mondal M, Roychoudhury S, Chakraborti S. Oxidant, mitochondria and calcium: an overview. Cell Signal 1999; 11 (2): 77-85.

[130] Duchen MR. Mitochondria and calcium: from cell signalling to cell death. J Physiol 2000; 529 Pt 1: 57-68.

1164   *Current Pharmaceutical Design,* 2005, *Vol. 11, No. 9*   *Plaetzer et al.*

[131]   Byrne AM, Lemasters JJ, Nieminen AL. Contribution of increased mitochondrial free Ca2+ to the mitochondrial permeability transition induced by tert-butylhydroperoxide in rat hepatocytes. Hepatology 1999; 29 (5): 1523-31.

[132]   Chernyak BV, Dedov VN, Chernyak V. Ca(2+)-triggered membrane permeability transition in deenergized mitochondria from rat liver. FEBS Lett 1995; 365 (1): 75-8.

[133]   Kowaltowski AJ, Castilho RF. Ca2+ acting at the external side of the inner mitochondrial membrane can stimulate mitochondrial permeability transition induced by phenylarsine oxide. Biochim Biophys Acta 1997; 1322 (2-3): 221-9.

[134]   Kowaltowski AJ, Castilho RF, Vercesi AE. Ca(2+)-induced mitochondrial membrane permeabilization: role of coenzyme Q redox state. Am J Physiol 1995; 269 (1 Pt 1): C141-7.

[135]   Kowaltowski AJ, Castilho RF, Vercesi AE. Opening of the mitochondrial permeability transition pore by uncoupling or inorganic phosphate in the presence of Ca2+ is dependent on mitochondrial-generated reactive oxygen species. FEBS Lett 1996; 378 (2): 150-2.

[136]   Xue LY, Chiu SM, Oleinick NL. Photochemical destruction of the Bcl-2 oncoprotein during photodynamic therapy with the phthalocyanine photosensitizer Pc 4. Oncogene 2001; 20 (26): 3420-7.

[137]   Usuda J, Azizuddin K, Chiu SM, Oleinick NL. Association between the photodynamic loss of Bcl-2 and the sensitivity to apoptosis caused by phthalocyanine photodynamic therapy. Photochem Photobiol 2003; 78 (1): 1-8.

[138]   Usuda J, Chiu SM, Murphy ES, Lam M, Nieminen AL, Oleinick NL. Domain-dependent photodamage to Bcl-2. A membrane anchorage region is needed to form the target of phthalocyanine photosensitization. J Biol Chem 2003; 278 (3): 2021-9.

[139]   Vantieghem A, et al. Phosphorylation of Bcl-2 in G2/M phase-arrested cells following photodynamic therapy with hypericin involves a CDK1-mediated signal and delays the onset of apoptosis. J Biol Chem 2002; 277 (40): 37718-31.

[140]   Deng X, Kornblau SM, Ruvolo PP, May WS, Jr. Regulation of Bcl2 phosphorylation and potential significance for leukemic cell chemoresistance. J Natl Cancer Inst Monogr 2001; (28): 30-7.

[141]   Hubmer A, Hermann A, Uberriegler K, Krammer B. Role of calcium in photodynamically induced cell damage of human fibroblasts. Photochem Photobiol 1996; 64 (1): 211-5.

[142]   Tajiri H, Hayakawa A, Matsumoto Y, Yokoyama I, Yoshida S. Changes in intracellular Ca2+ concentrations related to PDT-induced apoptosis in photosensitized human cancer cells. Cancer Lett 1998; 128 (2): 205-10.

[143]   Inanami O, Yoshito A, Takahashi K, Hiraoka W, Kuwabara M. Effects of BAPTA-AM and forskolin on apoptosis and cytochrome c release in photosensitized Chinese hamster V79 cells. Photochem Photobiol 1999; 70 (4): 650-5.

[144]   Ruck A, Heckelsmiller K, Kaufmann R, Grossman N, Haseroth E, Akgun N. Light-induced apoptosis involves a defined sequence of cytoplasmic and nuclear calcium release in AlPcS4-photosensitized rat bladder RR 1022 epithelial cells. Photochem Photobiol 2000; 72 (2): 210-6.

[145]   Li M, et al. Apoptosis induced by cadmium in human lymphoma U937 cells through Ca2+-calpain and caspase-mitochondria-dependent pathways. J Biol Chem 2000; 275 (50): 39702-9.

[146]   McGinnis KM, Gnegy ME, Park YH, Mukerjee N, Wang KK. Procaspase-3 and poly(ADP)ribose polymerase (PARP) are calpain substrates. Biochem Biophys Res Commun 1999; 263 (1): 94-9.

[147]   Blomgren K, et al. Synergistic activation of caspase-3 by m-calpain after neonatal hypoxia-ischemia: a mechanism of "pathological apoptosis"? J Biol Chem 2001; 276 (13): 10191-8.

[148]   Granville DJ, et al. Bcl-2 increases emptying of endoplasmic reticulum Ca2+ stores during photodynamic therapy-induced apoptosis. Cell Calcium 2001; 30 (5): 343-50.

[149]   Kessel D, Luo Y. Mitochondrial photodamage and PDT-induced apoptosis. J Photochem Photobiol B 1998; 42 (2): 89-95.

[150]   Hishita T, et al. Caspase-3 activation by lysosomal enzymes in cytochrome c-independent apoptosis in myelodysplastic syndrome-derived cell line P39. Cancer Res 2001; 61 (7): 2878-84.

[151]   Ishisaka R, et al. Participation of a cathepsin L-type protease in the activation of caspase-3. Cell Struct Funct 1999; 24 (6): 465-70.

[152]   Stoka V, et al. Lysosomal protease pathways to apoptosis. Cleavage of bid, not pro-caspases, is the most likely route. J Biol Chem 2001; 276 (5): 3149-57.

[153]   Lin CP, Lynch MC, Kochevar IE. Reactive oxidizing species produced near the plasma membrane induce apoptosis in bovine aorta endothelial cells. Exp Cell Res 2000; 259 (2): 351-9.

[154]   Skulachev VP. The p66shc protein: a mediator of the programmed death of an organism? IUBMB Life 2000; 49 (3): 177-80.

[155]   Fadok VA, Bratton DL, Rose DM, Pearson A, Ezekewitz RA, Henson PM. A receptor for phosphatidylserine-specific clearance of apoptotic cells. Nature 2000; 405 (6782): 85-90.

[156]   Ahmad N, Gupta S, Feyes DK, Mukhtar H. Involvement of Fas (APO-1/CD-95) during photodynamic-therapy-mediated apoptosis in human epidermoid carcinoma A431 cells. J Invest Dermatol 2000; 115 (6): 1041-6.

[157]   Yslas I, Alvarez MG, Marty C, Mori G, Durantini EN, Rivarola V. Expression of Fas antigen and apoptosis caused by 5,10,15, 20-tetra(4-methoxyphenyl)porphyrin (TMP) on carcinoma cells: implication for photodynamic therapy. Toxicology 2000; 149 (2-3): 69-74.

[158]   Azizuddin K, Kalka K, Chiu SM, Ahmad N, Mukhtar H, Separovic D. Recombinant human tumor necrosis factor alpha does not potentiate cell killing after photodynamic therapy with a silicon phthalocyanine in A431 human epidermoid carcinoma cells. Int J Oncol 2001; 18 (2): 411-5.

[159]   Klotz LO, Fritsch C, Briviba K, Tsacmacidis N, Schliess F, Sies H. Activation of JNK and p38 but not ERK MAP kinases in human skin cells by 5-aminolevulinate-photodynamic therapy. Cancer Res 1998; 58 (19): 4297-300.

[160]   Assefa Z, et al. The activation of the c-Jun N-terminal kinase and p38 mitogen-activated protein kinase signaling pathways protects HeLa cells from apoptosis following photodynamic therapy with hypericin. J Biol Chem 1999; 274 (13): 8788-96.

[161]   Huppertz B, Frank HG, Kaufmann P. The apoptosis cascade--morphological and immunohistochemical methods for its visualization. Anat Embryol (Berl) 1999; 200 (1): 1-18.

[162]   Motyl T, et al. Expression of bcl-2 and bax in TGF-beta 1-induced apoptosis of L1210 leukemic cells. Eur J Cell Biol 1998; 75 (4): 367-74.

[163]   Ormerod MG. Investigating the relationship between the cell cycle and apoptosis using flow cytometry. J Immunol Methods 2002; 265 (1-2): 73-80.

[164]   Ormerod MG, Sun XM, Brown D, Snowden RT, Cohen GM. Quantification of apoptosis and necrosis by flow cytometry. Acta Oncol 1993; 32 (4): 417-24.

[165]   Vermes I, Haanen C, Reutelingsperger C. Flow cytometry of apoptotic cell death. J Immunol Methods 2000; 243 (1-2): 167-90.

[166]   Walsh GM, Dewson G, Wardlaw AJ, Levi-Schaffer F, Moqbel R. A comparative study of different methods for the assessment of apoptosis and necrosis in human eosinophils. J Immunol Methods 1998; 217 (1-2): 153-63.

[167]   Sgonc R, Wick G. Methods for the detection of apoptosis. Int Arc Allergy Immunol 1994; 105 (4): 327-32.

[168]   Singh NP, McCoy MT, Tice RR, Schneider EL. A simple technique for quantitation of low levels of DNA damage in individual cells. Exp Cell Res 1988; 175 (1): 184-91.

[169]   O'Callaghan YC, Woods JA, O'Brien NM. Limitations of the single-cell gel electrophoresis assay to monitor apoptosis in U937 and HepG2 cells exposed to 7beta-hydroxycholesterol. Biochem Pharmacol 2001; 61 (10): 1217-26.

[170]   Negoescu A, et al. In situ apoptotic cell labeling by the TUNEL method: improvement and evaluation on cell preparations. J Histochem Cytochem 1996; 44 (9): 959-68.

[171]   Colecchia M, et al. Detection of apoptosis by the TUNEL technique in clinically localised prostatic cancer before and after combined endocrine therapy. J Clin Pathol 1997; 50 (5): 384-8.

[172]   Labat-Moleur F, et al. TUNEL apoptotic cell detection in tissue sections: critical evaluation and improvement critical evaluation and improvement. J Histochem Cytochem 1998; 46 (3): 327-34.

[173]   Whiteside G, Munglani R. TUNEL, Hoechst and immunohistochemistry triple-labelling: an improved method for detection of apoptosis in tissue sections--an update. Brain Res Brain Res Protoc 1998; 3 (1): 52-3.

[174]   Pulkkanen KJ, Laukkanen MO, Naarala J, Yla-Herttuala S. False-positive apoptosis signal in mouse kidney and liver detected with TUNEL assay. Apoptosis 2000; 5 (4): 329-33.

[175]   Stennicke HR, Salvesen GS. Caspase assays. Methods Enzymol 2000; 322: 91-100.

[176]   Gurtu V, Kain SR, Zhang G. Fluorometric and colorimetric detection of caspase activity associated with apoptosis. Anal Biochem 1997; 251 (1): 98-102.

[177]   Gorman AM, Hirt UA, Zhivotovsky B, Orrenius S, Ceccatelli S. Application of a fluorometric assay to detect caspase activity in thymus tissue undergoing apoptosis *in vivo*. J Immunol Methods 1999; 226 (1-2): 43-8.

[178]   Nicholson DW. Caspase structure, proteolytic substrates, and function during apoptotic cell death. Cell Death Differ 1999; 6 (11): 1028-42.

[179]   Mack A, Furmann C, Hacker G. Detection of caspase-activation in intact lymphoid cells using standard caspase substrates and inhibitors. J Immunol Methods 2000; 241 (1-2): 19-31.

[180]   Kohler C, Orrenius S, Zhivotovsky B. Evaluation of caspase activity in apoptotic cells. J Immunol Methods 2002; 265 (1-2): 97-110.

[181]   Vermes I, Haanen C, Steffens-Nakken H, Reutelingsperger C. A novel assay for apoptosis. Flow cytometric detection of phosphatidylserine expression on early apoptotic cells using fluorescein labelled Annexin V. J Immunol Methods 1995; 184 (1): 39-51.

[182]   Zhang G, Gurtu V, Kain SR, Yan G. Early detection of apoptosis using a fluorescent conjugate of annexin V. Biotechniques 1997; 23 (3): 525-31.

[183]   van Engeland M, Nieland LJ, Ramaekers FC, Schutte B, Reutelingsperger CP. Annexin V-affinity assay: a review on an apoptosis detection system based on phosphatidylserine exposure. Cytometry 1998; 31 (1): 1-9.

[184]   Bossy-Wetzel E, Green DR. Detection of apoptosis by annexin V labeling. Methods Enzymol 2000; 322: 15-8.

[185]   Bossy-Wetzel E, Green DR. Assays for cytochrome c release from mitochondria during apoptosis. Methods Enzymol 2000; 322: 235-42.

[186]   Lemasters JJ. V. Necrapoptosis and the mitochondrial permeability transition: shared pathways to necrosis and apoptosis. Am J Physiol 1999; 276 (1 Pt 1): G1-6.

[187]   Mosmann T. Rapid colorimetric assay for cellular growth and survival: application to proliferation and cytotoxicity assays. J Immunol Methods 1983; 65 (1-2): 55-63.

[188]   Reers M, Smith TW, Chen LB. J-aggregate formation of a carbocyanine as a quantitative fluorescent indicator of membrane potential. Biochemistry 1991; 30 (18): 4480-6.

[189]   Smiley ST, *et al.* Intracellular heterogeneity in mitochondrial membrane potentials revealed by a J-aggregate-forming lipophilic cation JC-1. Proc Natl Acad Sci USA 1991; 88 (9): 3671-5.

[190]   Salvioli S, Ardizzoni A, Franceschi C, Cossarizza A. JC-1, but not DiOC6(3) or rhodamine 123, is a reliable fluorescent probe to assess Delta Psi changes in intact cells: Implications for studies on mitochondrial functionality during apoptosis. FEBS Lett 1997; 411 (1): 77-82.

[191]   Bernardi P, Scorrano L, Colonna R, Petronilli V, Di Lisa F. Mitochondria and cell death. Mechanistic aspects and methodological issues. Eur J Biochem 1999; 264 (3): 687-701.

[192]   Richter C, Schweizer M, Cossarizza A, Franceschi C. Control of apoptosis by the cellular ATP level. FEBS Lett 1996, 378 (2): 107-10.

[193]   Neufeld HA, Towner RD, Pace J. A rapid method for determining ATP by the firefly luciferin-luciferase system. Experientia 1975; 31 (3): 391-2.

[194]   Kiesslich T, Oberdanner CB, Krammer B, Plaetzer K. Fast and reliable determination of intracellular ATP from cells cultured in 96-well microplates. J Biochem Biophys Methods 2003; 57 (3): 247-51.

[195]   Bergmann-Leitner ES, Duncan EH, Leitner WW. Identification and targeting of tumor escape mechanisms: a new hope for cancer therapy? Curr Pharm Design 2003; 9(25): 2009-23.

[196]   Pellecchia M, Reed JC. Inhibition of anti-apoptotic Bcl-2 family proteins by natural polyphenols: new avenues for cancer chemoprevention and chemotherapy. Curr Pharm Design 2004; 10(12): 1387-98.

[197]   Schaffer M, Ertl-Wagner B, Schaffer PM, Kulka U, Hofstetter A, Duhmke E, *et al.* Porphyrins as radiosensitizing agents for solid neoplasms. Curr Pharm Design 2003; 9(25): 2024-35.

(ii) List of Abbreviations

αGDH	α-glycerophosphate dehydrogenase
ACO	aconitase
AIF	apoptosis inducing factor
AlPcS$_4$	aluminum (III) phthalocyanine tetrasulfonate
ANT	adenine nucleotide translocator
Apaf-1	apoptotic protease-activating factor-1
BCA	bicinchoninic acid
BCNU	1,3-bis-(2-chlorethyl)-1-nitrosurea
Bis	bisindolylmaleimide
BPR	benzodiazepine peripheral receptor
BSO	DL-buthioninesulfoximine
C-I – C-IV	complex I to IV of the mitochndrial respiratory chain
CAD	caspase-activated DNase
Cat	catalase
CDDP	cis-dichlorodiammineplatinum, cisplatin
c-H$_2$DCFDA	5-(and 6-)-carboxy-2',7'-dichlorodihydrofluorescein diacetate
CK	creatine kinase
CoQ(H$_2$) / Q	reduced form of coenzyme Q, also ubiquinone
COX	cytochrome c oxidase
Cph.D	cycophilin-D
Cyt.c / c	cytochrome c
Cyt.b / b	cytochrome b
ΔΨ	mitochondrial transmembrane potential
dATP	2'-deoxyadenosin 5'triphosphate
DCF	2'-7'-dichlorofluorescein
(H$_2$)DCFDA	2',7'-dichlorodihydrofluorescein diacetate
DEVD-AMC	7-amino-4-methyl-coumarin
DHOH	Dihydroorotate dehydrogenase
DIABLO	direct IAP-binding protein with low pI
DISC	death-inducing signaling complex
DlD	dihydrolipoaminde dehydrogenase
DMEM	Dulbecco's modified Eagle's medium

DMSO	dimethylsulfoxide
DTT	dithiotreitol
EDTA	ethylene diamine tetra-acetate
ER	endoplasmatic reticulum
FADD	Fas-associated death domain protein
$FADH_2$	reduced form of flavin adenine dinucleotide
FAD^+	oxidzed form of flavin adenine dinucleotide
FCS	fetal calf serum
FET	forward electron transfer
FMN	flavin mononucleotide
GPX	glutathione peroxidase
GR	glutathione reductase
$Grx2_{red}$	glutaredoxin-2 reduced
$Grx2_{ox}$	glutaredoxin-2 oxidized
GSH	reduced form of glutathione
GSH-ee	Glutathione monoethyl ester
GSSG	oxidized form of glutathione
GST	Glutathione-S-transferase
HEPES	2-[4-(2-hydroxyethyl)-1-piperazinyl] ethanesulfonic acid
HK	hexokinase
IAP	inhibitor of apoptosis
ICE	interleukin-1β converting enzyme
IDH1	isocitric dehydrogenase, $NADP^+$-dependent
IDH	isocitric dehydrogenase, NAD^+-dependent
IL-1β	interleukin-1β
JC-1	5,5', 6,6'-tetrachloro- 1,1', 3,3'-tetraethyl benzimidazolyl carbocyanine iodide
JNK	c-Jun N-terminal kinase
KGDHC	α-Ketoglutarate dehydrogenase complex
LPS	lipopolysaccharide
MAOs	Monoamine oxidases
MDH	malate dehydrogenase
ME	malic enzyme, $NADP^+$-dependent

MnSOD	mitochondrial manganese superoxide dismutase
mPTP	mitochondrial permeability transition pore
MTT	3-[4,5-dimethylthiazol-2-yl]-2,5-diphenyl tetrazolium bromide
NAD(P)H	reduced form of nicotinamide adenine dinucleotide (phosphate)
NAD(P)$^+$	oxidized form of nicotinamide adenine dinucleotide (phosphate)
PBS	phosphate-buffered saline
PDHC	pyruvate dehydrogenase complex
PDT	photodynamic treatment
PGPX	phospholipid hydroperoxide glutathione peroxidase
PKC	protein kinase C
P_i	inorganic phosphate
PPP	pentosephosphate pathway
PCD	programmed cell death
$Prx3_{red}$	peroxiredoxin reduced
Prx_{ox}	peroxiredoxin oxidized
RET	reverse electron transfer
ROS	reactive oxygen species
SDH	succinate dehydrogenase
Smac	second mitochondrial-derived activator of caspases
SOD	superoxide dismutase
STS	staurosporine
TCA	tricarboxylic acid cycle, citric acid cycle
TH	transhydrogenase
TNF	tumor necrosis factor
Trx	thioredoxin
$Trx2_{red}$	thioredoxin-2 reduced
$Trx2_{ox}$	thioredoxin-2 oxidized
TrxR2	thioredoxin-2 reductase
VDAC	voltage dependent anion channel

(iii) List of Tables, Figures and Reactions

www.ingramcontent.com/pod-product-compliance
Lightning Source LLC
Chambersburg PA
CBHW070737220326
41598CB00024BA/3452